Digital Atlas of Traditional Foods
Made from Cereals and Milk

Groningen Archaeological Studies Volume 33

Groningen Institute
of Archaeology
Poststraat 6
9712 ER Groningen
the Netherlands
gia@rug.nl

Website
www.plantatlas.eu

Publisher's address
Barkhuis
Kooiweg 38
9761 GL Eelde
the Netherlands
info@barkhuis.nl
www.barkhuis.nl

Digital Atlas of Traditional Food Made from Cereals and Milk

R.T.J. Cappers

Barkhuis &
University of Groningen Library
Groningen 2018

Photos: René T. J. Cappers (RC), Dirk Fennema (DF), Hussein Madina (HM),
Reinder Neef (RN) and Yasar Okur (YO).

Book cover design: Ebel Kuipers
Book interior design and typesetting: Ebel Kuipers

Copy editor: Suzanne Needs-Howarth

Printed by: KOPA, Kaunas, Lithuania

Cover illustration: René T. J. Cappers

ISBN 978-94-92444-70-7

Contents

ATLAS *105*

Preface

Every year, in various parts of the world, people harvest barley and wheat from fields and process the grain into food. The processing of the harvest and food preparation that take place today are based on an experience of around 14 000 yearly harvests, starting in south-west Asia. The first 4000 yearly harvests involved wild grains, whereas subsequent harvests involved domesticated cereals. Once domesticated cereals started to be harvested and processed, it took another 1000 years before pottery became available and another 1700 years before metal utensils, such as containers, became available. At the same time as the first crops became domesticated, the first animals were being domesticated. The presence of milk residues on early pottery suggests that milk became part of the human diet quite early on as well, although some 2000 years after the first domestication of crops. This implies that late hunters-gatherers and early farmers had a long trial period during which they had the opportunity to gain experience and develop technologies for the efficient processing of grain and milk.

Initially, food production was dependent on plants and animals from the local environment. Through trade contacts, new '*ingredients*' gradually became available. The traditional '*cuisine*' that characterizes the cultural identity of a certain region is therefore quite changeable and often reflects a tradition covering only a limited period of time. The preservation of traditional cuisines can be explained by the availability of stable food sources and the advanced technology that is necessary for '*harvesting*' and milking, for post-harvesting processing and milk processing, and for final food preparation. This technology is based on basic processes that play a role in cereal and milk processing and food preparation, as well as other crops.

The examples of traditional foods that are presented here most probably have a long tradition. Some of them even go back to the transition from hunting and gathering to early agriculture. These traditional food items are characterized by their amenability to large-scale production and by their long shelf-life, both of which enabled efficient use of labour and fuel. Examples of such foods that were recently collected in south-west Asia and North Africa are presented in the atlas section of this book. These examples clearly illustrate the variation that can be produced even when food is based on simple recipes.

Direct evidence of the characteristics of traditional foods can be obtained from the analysis of subfossil food remains. Some recent publications show an increased interest in the study of such food remains and suggest updates to the standards for their documentation (e.g. Carretero et al., 2017; Heiss et al., 2017; Samuel, 2000). Other recent publications argue that archaeobotanical research and archaeozoological research should become more integrated and more closely linked to the material culture related to crop and milk processing and food preparation, such as pottery, implements for *'grinding'* and *'pounding'*, and heating installations (Cappers et al., 2016; Vigne, 2015).

In order to assess the quality of traditional foods, further research is required. It is important, for example, to achieve a taxonomic identification of the grain down to the level of subspecies and to determine which processes have been applied to the grain. This is particularly relevant when these processes cause reallocation of nutrients, such as in *'par-boiling'*, or when a part of the nutrients is removed, such as in *'polishing'* and *'pearling'*.

All of the traditional foods documented in this book have been archived in the reference collection of the Groningen Institute of Archaeology (GIA) and are available for such research. This will further increase our knowledge of traditional foods and of the criteria that played a role in crop selection, which in ancient times were often different than in modern times.

I would like to thank everyone who helped me during my fieldwork and in collecting and documenting the samples of traditional foods.

René Cappers
Groningen, May 2018

1
Introduction

1.1 Late foragers and early farmers

1.1.1 Landscape exploitation

For most of their history, humans have been supplied with food by means of hunting and gathering. It is only recently that agriculture was developed as a new form of food economy. Agriculture is characterized by the cultivation of crops in fields and the control and protection of animals, during which these animals are also partly housed in or near the settlement.

For both hunter-gatherers and farmers, the landscape must provide enough food for them to survive. There must also be sufficient fuel available for food preparation. Both plant material and animal dung can be used as fuel. When plant material is used as fuel, it is desirable to only use dead plant parts, such as the dead matter of herbaceous plants and dead branches of shrubs and trees. Such dead plant material has the advantage that it burns well and that its use does not affect the carrying capacity of the vegetation as a supplier of fuel.

Food is necessary for the building, repair, and maintenance of body tissues and as an energy source. In order for an organism to survive, the energy balance will have to be positive. The amount of energy needed for collecting or producing food, for collecting fuel, and for preparing food should be at least equal to the amount of energy needed to stay alive.

The energy present in items of food is ultimately all fixed by green plants. This energy originates from sunlight and is used by green plants to make sugar from water and carbon dioxide. Green plants thus form the first link in every food chain. During the last ice age, the availability of water was a limiting factor for plant growth, and therefore also for other organisms, including humans. This forced humans to hunt and gather in small groups in relatively large areas. Research at the 23 000-year-old settlement of Ohalo II, located on the shore of the Sea of Galilee, shows that former hunter-gatherers in this area still had enough plants available to them to survive and that they were able to feed on a wide spectrum of plants.

Many temporary camp sites and settlements of hunter-gatherers have been identified in the eastern Mediterranean region dating from the final phase of the last glaciation, the Late Pleistocene (12 700–9 600 BCE). Apparently the conditions were sufficiently favourable for these hunter-gatherers (referred to as Natufians) that they were able to keep themselves alive in small groups. The food collected by the Natufians consisted in part of wild grains and wild legumes. During the Younger Dryas (10 800–9 700 BCE), the climate deteriorated and was once again characterized by Ice Age conditions, and hunter-gatherers were once again challenged to adapt to changing vegetation. Pollen research at various locations in south-west Asia has shown that the climate became warmer and more humid some 10 000 years ago (Zeist & Bottema, 1991). This climatic improvement marks the end of the last ice age, and with that the end of the Pleistocene and the beginning of the current geological epoch, which is called the Holocene. Of special interest in relation to the early food economy is the development of a so-called Mediterranean climate, which is characterized by cool and wet winters and warm and dry summers. This climate extended, and still extends, over large parts of the Mediterranean basin and the area contiguous with the Taurus Mountains and the Zagros Mountains.

As a result of this climatic improvement, herbaceous plants and trees expanded from their glacial refugia. Trees need more water than do herbaceous plants, and their expansion in particular is indicative of the increased availability of water. Different types of vegetation developed, including a tree steppe with undergrowth that was sometimes dominated by huge populations of the progenitors of the founder crops of agriculture. Characteristic trees were deciduous Oak species (including *Quercus cerris*, *Q. ithaburensis*, and *Q. robur*), Almond (*Prunus dulcis*), and Pistachio species (including *Pistacia terebinthus* and *Pistacia atlantica*).

Although most of that landscape has now been radically changed through the impact of humans, some large populations of wild cereals are still present where grazing is prevented (figure 1; see also Anderson 1999: figure 12.24). Similarly large concentrations of wild food plants must have become attractive for hunter-gatherers to harvest on a large scale. But to be able to make use of such large quantities of grain and '*seeds*', which were only available seasonally, they would have to have been stored.

FIGURE 1: A population of wild Barley (*Hordeum vulgare* ssp. *spontaneum*) in south-eastern Anatolia. Most of the plants are present without their spikes (Susuz Dagleri, Turkey; October 1998; RN).

The desire to make use of these wild foods, in turn, would have necessitated the protection of the stored harvest, which will have been an important stimulus for hunter-gatherers to give up their mobile lifestyle for a temporary or more permanent sedentary lifestyle. Suitable locations for settlements will have been chosen on the basis of their strategic location in terms of defence and proximity to water. If such suitable locations were at a considerable distance from fields with wild grains and wild pulses, it would have been desirable for hunter-gatherers to create fields in the vicinity of the settlement. In order for them to create fields closer to the settlement, hunter-gatherers would have to have been aware of the life cycle of plants, specifically the fact that a seed that is buried in the soil will provide a new generation the next year. In this way, protection of crop plants could be optimized, transport of harvests over long distances was no longer necessary, and the harvest could be safely stored and well protected.

Contemporaneous with the development of the first settlements and the creation of fields, the first animals were also integrated with the *domus* (home) of humans, receiving care and protection (Zeder, 2011). In west Asia, sheep, goat, cattle, and pig are the first animals to become domesticated. These animals provide usable products both during their lifetime and after their death (figure 2).

		goat	sheep	cattle	pig
traction				○	
skin	hair	○	○		
	leather	□	□	□	□
milk		○	○	○	
dung				○	
meat		□	□	□	□
bones		□	□	□	□

FIGURE 2: Main products obtained from goat, sheep, cattle, and pig (○= ante-mortem products; □ = post-mortem products).

By integrating crop production and keeping domestic animals, humans created a mixed agricultural economy in which a large part of the landscape could be, and continues to be, exploited in a controlled manner. The first trophic level, represented by plants, offers a large carrying capacity of energy. It is obtained from a combination of fields, grasslands, and vegetation where edible plant parts are collected from wild-growing plants and where hunting is still practiced. Humans are thus a consumer feeding on two trophic levels: as plant eaters and as eaters of animal products.

In the traditional agricultural systems of the Mediterranean basin, south-west Asia, and North Africa that are the focus of this book, the crops cultivated in the fields are generally intended for humans, although in exceptional cases, crop plants are given to animals. For example, grain can be given as an extra food supply to a cow prior to hard work, such as ploughing a field. It can also happen that the quality of a grain harvest is poor and that it would take too much labour to remove the impurities. The grain stock can then be downgraded to animal feed. Sufficient amounts of hay will have to be harvested and stored as winter feed for sheep, goats, and cattle. In some cases, some members of the herd are slaughtered before the onset of the season during which the available vegetation can no longer support the entire herd.

Sheep, goats, and cattle thus feed primarily on the vegetation growing outside the fields. The arable weeds that are a companion of crop plants can only be grazed once the harvest has been removed from the field (figures 3 and 4). While sheep and cattle are grazers and feed on herbaceous plants close the ground and on branches of trees and shrubs that are accessible to them by stretching their necks, goats are browsers, and they can also climb trees. Goats thus have access to more plant material than do sheep and cattle. Pigs are real omnivores and feed on a large variety of organic material that is available both in the subsoil and on the ground surface, including kitchen waste (figure 5). Animal dung from animals that can be easily collected and shaped can subsequently be dried and used as fuel. It is also possible to use dung as a fertilizer when long-term harvesting has depleted the minerals in the soil.

FIGURE 3: This landscape is providing humans with food on two levels. The flat land in the foreground, which can be ploughed without the risk of erosion, is used for crop production. The cereal crop has been recently harvested. The vegetation on the slopes is grazed by sheep and goats and will nourish farmers with milk and meat (west of Kanimaran, Iraq; June 2014; RC).

FIGURE 4: After harvesting of the cereal crop, sheep and goats are allowed to graze the stubble field. The culms of cereals are cut at some distance above the ground to prevent damage to the cutting tool. The low-growing arable weeds offer good fodder and are partly converted into milk and meat (north of Dagalah, Iraq; August 2013; RC).

FIGURE 5: Browsing of trash that is concentrated at the edge of a street by goats and a pig. Discarded low-quality and inedible food is recycled into high-quality food, namely, milk and meat (Kharagpur, India; September 2014; RC).

1.1.2 Cereal domestication

1.1.2.1 *The concept of domestication*

Plants that are exploited by humans can be designated as '*economic plants*'. These plants are distinguished from '*wild plants*', which are not used by humans. Economic plants can be used, for example, as food, as medicine, as fuel, for colouring and tanning, as construction material for making basketry or shelter, and to obtain fibres or hairs for the production of fabrics.

Within the economic plants, a further distinction is made here between '*cultivated plants*' and '*domesticated plants*'. In a cultivated plant, the plant is still capable of '*seed dispersal*' and thus of propagating independently of humans. In a domesticated plant, in contrast, the seed distribution is blocked (Harlan, 1992). Note that the term '*seed dispersal*' is used regardless of the composition of the dispersal unit (also known as '*diaspore*'). The dispersal unit of a plant can be a seed; a '*fruit*' containing one or more seeds; or one or more fruits surrounded by vegetative plant parts (exocarp). The dispersal unit of a grass is a 1-seeded fruit with its vegetative plant parts or, in the case of some domestic grasses, only the 1-seeded fruit ('*grain kernel*').

The majority of the economic plants are cultivated, whereas only a limited number have become domesticated. An important group of domesticated plants is represented by cereals and pulses. The first cereals to be domesticated were barley and wheat. The exploitation of barley and wheat initially involved wild cereals, which were still characterized by natural seed dispersal. This initial phase of exploitation is referred to as pre-domestication cultivation (for a recent discussion, see Asouti & Fuller, 2013).

The domestication of barley and wheat involved not only the hampering of seed dispersal, but also a complex of changes in both morphological and physiological characteristics (figure 6). Hammer (1984) introduced the term '*domestication syndrome*' to refer to diseases that are characterized by a complex of symptoms. The hampering of seed dispersal is considered to be a major change. For plants that cannot propagate vegetatively, this change is lethal. Of course to us humans, the reality of a farmer sowing a field is more likely to suggest a successful symbiosis than a complex of changes that recall a disease. After all, there is mutual dependency: the plant has become dependent on humans, who use part of the harvest for sowing in order to allow seeds to germinate and to produce a next generation. And humans have become more and more dependent for their food supply on these domesticated crops. The success of this symbiosis is supported by the current large-scale cultivation of domesticated crops, even outside their original distribution area.

Characteristics	**Morphological traits**	**Physiological traits**
Growing season		
– loss of vernalization		○
– different photoperiodic response		○
Yield		
– reduced dormancy		○
– increased photosynthesis		○
– increased tillering	○	
– increased number of grain kernels	○	
– increased grain size	○	
Harvesting and crop processing		
– synchronous fruit ripening		○
– reduced rachis brittleness	○	
– reduced hull tightness	○	
– reduced appendages (awns and hairs)	○	

FIGURE 6: Morphological and physiological domestication traits of cereals categorised in relation to growing season, yield, and harvesting and crop processing.

The domestication of barley and wheat is a complex process in which selection by humans was initially unconscious. The morphological and physiological changes have a genetic basis and are caused by mutations. The domestication of barley involved mutations and selection within a single species. The domestic history of wheat is more complex because hybridization has also taken place between different types of Wheat (*Triticum*) and Goat beard (*Aegilops*), a related grass genus.

1.1.2.2 *Domestication and taxonomy*

The large variety within domesticated barley and wheat is taxonomically classified into subspecies. In barley we are dealing with three subspecies: wild Barley (*Hordeum vulgare* ssp. *spontaneum*), domesticated 2-row Barley (*Hordeum vulgare* ssp. *distichon*), and domesticated 6-row Barley (*Hordeum vulgare* ssp. *vulgare*). In barley, the classification on the subspecies level is based on the '*rachis brittleness*' and on the fertility of the '*spikelets*'. In the case of 2-row barley, only the central spikelet is fertile, while in the case of 6-row barley, the two lateral spikelets are also fertile (figure 7). Both domesticated subspecies have become available as '*hulled*' and '*naked*' forms, but this difference in '*hull tightness*' is not taken into account in the taxonomic classification.

species	wild subspecies	domesticated subspecies	
	brittle rachis	non-brittle rachis	
	reduced fertility (2-row)		full fertility (6-row)
H. vulgare	ssp. *spontaneum*	ssp. *distichon*	ssp. *vulgare*

FIGURE 7: Modern classification of wild and domesticated Barley (*Hordeum vulgare*). The classification takes into account the rachis brittleness and the fertility of the spikelets. The first column presents the species level (H. = *Hordeum*), and the next three columns represent the subspecies (ssp.) levels. The green shading indicates the ploidy level (: diploid taxa).

The domestication of wheat involved six species of Wheat (*Triticum*) and two species of Goat beard (*Aegilops*). As a result of hybridization between Wheat (*Triticum*) and Goat beard (*Aegilops*), the number of chromosomes per cell nucleus varies among the different wheat species. Two wheat species are diploid (meaning they have one pair of chromosomes per cell nucleus), two wheat species are tetraploid (meaning they have two pairs of chromosomes per cell nucleus), and two wheat species are hexaploid (meaning they have three pairs of chromosomes per cell nucleus). The presence of many copies of genes contributes to the large variability in the phenotypic expression of all kinds of features and has resulted in the development of many cultivars. In wheat, the classification into species and subspecies is based on differences in the rachis brittleness and the hull tightness. In the case of wheat, in contrast to barley, the domesticated subspecies is indicative of both the rachis brittleness (being the primary domestication feature) and the hull tightness (being the secondary domestication feature) (figure 8).

species	wild subspecies	domesticated subspecies	
	brittle rachis	semi-brittle rachis	non-brittle rachis
	hulled grain		naked grain
T. aestivum		ssp. *spelta*	ssp. *aestivum*
		ssp. *macha*	ssp. *compactum*
		ssp. *vavilovii*	ssp. *sphaerococcum*
T. zhukovskyi		(only domesticated)	
T. turgidum	ssp. *dicoccoides*	ssp. *dicoccon*	ssp. *turgidum*
		ssp. *paleocolchicum*	ssp. *durum*
			ssp. *carthlicum*
			ssp. *polonicum*
			ssp. *turanicum*
T. timopheevii	ssp. *armeniacum*	ssp. *timopheevii*	
T. monococcum	ssp. *aegilopoides*	ssp. *monococcum*	
T. urartu	(only wild)		

FIGURE 8: Modern classification of wild and domesticated Wheat (*Triticum*). The classification takes into account the rachis brittleness and the hull tightness. The first column presents the species level (T. = *Triticum*), and the next three columns represent the subspecies (ssp.) levels. The green shading indicates the ploidy level : diploid taxa; : tetraploid taxa; : hexaploid taxa.

1.1.2.3 *Different phases of domestication*

The domestication process of barley and wheat is closely intertwined with harvesting and crop processing. It is characterized by two morphological features that are involved in seed dispersal: the rachis brittleness and the hull tightness. The rachis brittleness can be considered as a primary domestication feature because it has resulted in the blocking of the seed dispersal. The hull tightness can be considered as a secondary domestication characteristic because it makes seed dispersal possible once more, due to a change in the dispersal unit.

In wild barley and wild wheat, seed dispersal is possible because the '*rachis*' (alternative spelling: '*rhachis*') will fragment when the grains are ripe. Within a '*spike*', the spikelets (each containing one or more grain kernels) gradually ripen, from top to bottom. Once the grain kernels are ripe, the rachis breaks underneath the spikelet, where the internode is attached to the subsequent node, and dissemination takes place. The dispersal unit in wild barley consists of a '*rachis segment*' that supports three spikelets. The dispersal unit in wild wheat consists of a rachis segment that supports a single spikelet (see section 1.2).

The first domesticated barley and wheat were hulled, which means that the grain kernels were still connected with '*chaff*' after '*threshing*', similar to the dispersal units of their wild progenitors. In a later stage of crop domestication, forms of barley and wheat appeared in which the '*hulls*' are only loosely wrapped around the grain kernels. Such forms are called '*naked*' because both the dispersal unit and the threshing unit are a grain kernel without chaff (see also the discussion on '*free-threshing*' in the Glossary). Because seed dispersal is once again possible in these naked cereals, such cereals are no longer fully domesticated in the sense that Harlan (1992) defined the term.

The exploitation of barley and wheat shows different phases throughout the domestication process, and these phases influence the final food preparation. Four properties of the grain are of particular interest: the grain size, the ripeness of the grain kernels, the rachis brittleness, and the hull tightness. Ripe grain has the advantage that the moisture content is low, meaning that it can be stored in good condition for a long time. Ripe grain can also be used for sowing (see figure 44 for a visualization of how to obtain seeds for sowing).

In the first phase, only the ripe dispersal units of barley and wheat are collected (figure 9: 1). In the case of wild barley, '*rachillate triplets*' are collected, but only the central spikelet of such a rachillate triplet contains a grain kernel. And in wild wheat, '*rachillate spikelets*' are collected that contain, depending on the fertility, one or two grain kernels. Fragmentation of these '*harvesting units*' results in a '*floret*' in barley and in a grain kernel in wheat. '*De-graining*' (also termed '*de-hulling*', '*hulling*', '*de-husking*' or '*husking*') of the rachillate spikelet is necessary because the inedible vegetative plant parts are substantial. The need to further process the floret of barley is dependent on the grain size. Initially, the grain kernel of wild barley was rather small, and further fragmentation may have been desirable to get rid of the firm hulls. Sowing the rachillate triplets in deep soil layers, reducing competition from arable weeds, and improved crop cleaning will have resulted in an increase in the grain size, thereby reducing the desirability of removing the hulls. Measurements of subfossil grain kernels indicate that this increase of grain size came about during the Pre-Pottery Neolithic (Fuller et al., 2010).

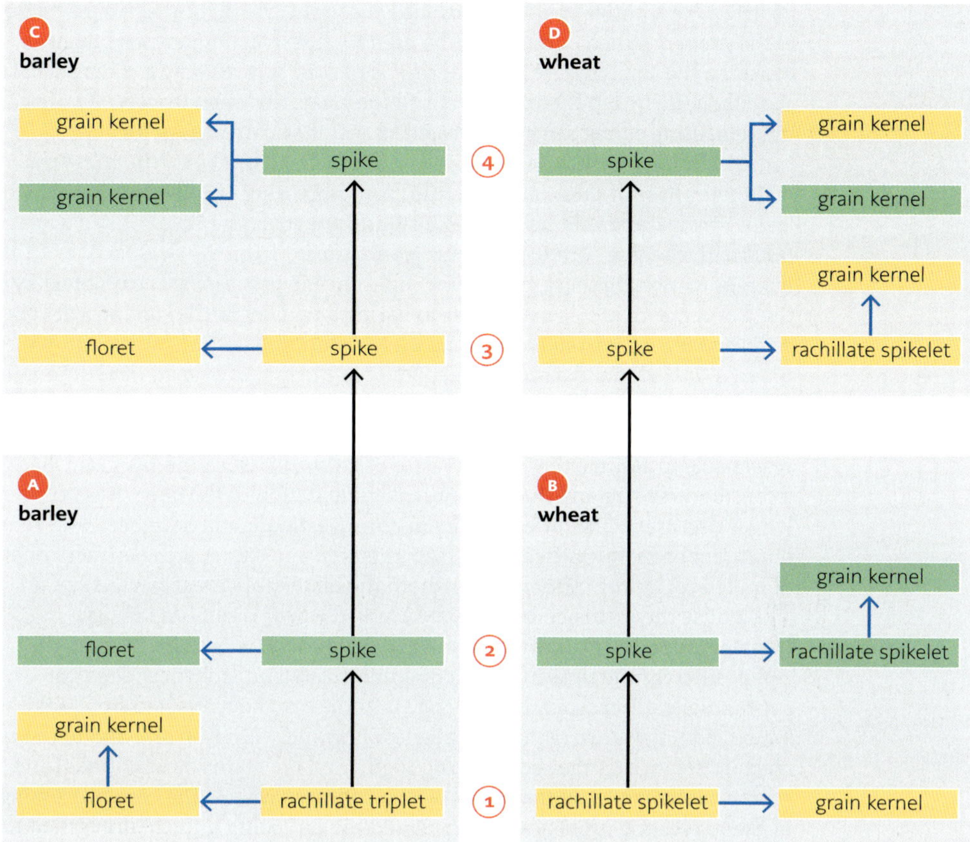

FIGURE 9: Four phases in the development of harvesting (black arrow) and further processing (blue arrow) of wild (A and B) and domesticated (C and D) barley and wheat into edible units. Fully ripe units are presented in yellow and partly unripe units, in green. 1: Harvesting and processing of **ripe** upper parts of the spikes of **wild** cereals. 2: Harvesting and processing of **partly unripe** spikes of **wild** cereals. 3: Harvesting and processing of **fully ripe** spikes of **hulled** cereals with **non-brittle rachis** (3C) or **semi-brittle rachis** (3D) (= fully domesticated). 4: Harvesting and processing of **partly unripe** spikes of **naked** cereals with **non-brittle** rachis (= semi-domesticated) and further processing of either partly unripe or fully ripe grain.

The harvest of wild grain can be maximized in two ways, by harvesting several times from the same population (figure 9: phase 1), or by harvesting intact spikes as soon as the ripening process has started (figure 9: phase 2). In the latter case, some of the grain kernels were eaten while they were still unripe. Harvesting spikes that are not yet fully ripe requires a '*harvesting knife*' or sickle to cut the non-brittle culm. There are two options for obtaining seeds for sowing. They can be collected in a first harvesting phase, similar to phase 1 in figure 9. Alternatively, a sufficient quantity of ripe rachillate triplets (for barley) or rachillate spikelets (for wheat) can be isolated from the partially ripe spikes before further crop processing is undertaken.

Once humans started to cultivate cereals in which the rachis was no longer fragmented, it was possible to harvest complete spikes that only contain ripe grain kernels (figure 9, phase 3). When culms and leaves are to serve as fuel for *'roasting'* (alternative spelling: *'toasting'*) the grain, the plants have to be cut low to the ground. When culms are to be used for construction purpose, such as basketry, the ripe, complete spikes can be harvested with only a short part of the culm attached to them, to obtain all the grain, and then in a second phase, the remaining culms can be harvested so that they can be used for construction purposes.

In this phase of domestication, the crop processing of hulled barley became less laborious, and this gave barley a significant advantage over hulled wheat (alternative but not preferred term: *'glume wheat'*) from the perspective of the humans exploiting these grains. The difference in processing is explained by the difference in rachis brittleness in barley (viz. non-brittle) and wheat (viz. semi-brittle), which results in completely different threshing units: a floret in the case of barley and a rachillate spikelet in the case of wheat. In addition, the difference in hull tightness and the different construction of the spike are also determinant in the design of these threshing units (figure 37). It seems reasonable to assume that in this phase of crop domestication the grain was of a considerable size and that therefore fragmentation of the floret was no longer necessary. The difference in labour investment will have been an important economic criterion in crop selection and explains the attractiveness of hulled barley for farmers throughout the history of cereal cultivation.

Because the rachis of domesticated hulled wheat is semi-brittle, its fragmentation by threshing is similar to that of wild wheat. As a result, the composition of the threshing unit of domesticated hulled wheat does not substantially differ from the dispersal unit that can be harvested from wild wheat. Both units are rachillate spikelets. They only differ in size and in the fracture surface on the rachis, being regular on the rachis of wild wheat and somewhat irregular on the rachis of domesticated wheat. The crop processing of the harvesting units of wild wheat and the threshing units of domesticated hulled wheat does not differ. In both cases, it includes the isolation of the grain kernels from their vegetative plant parts. This is necessary because the amount of vegetative plant parts is relatively large in comparison to the accompanying grain kernels. Moreover, the hull tightness is less in comparison with hulled barley. This will cause uneven fragmentation if whole rachillate spikelets are granulated or pulverized. Separation of edible and non-edible after granulated or pulverized is therefore troublesome and will result in the loss of part of the fragmented grain.

The rachis of domesticated hulled barley is tough, and threshing results in an irregular fragmentation of the rachis and the detachment of the florets from the spike. The threshing unit of domesticated hulled barley is therefore a floret. Because the hulls are almost fused with the grain kernels, *'granulation'* and pulverization of the entire threshing unit (floret) is possible. This means that the crop processing and food preparation of hulled barley is much less labour intensive than that of hulled wheat.

Once naked cereals became available, at a later stage of the domestication process, the harvesting method had to be adjusted to obtain a maximum harvest (figure 9: phase 4). This adjustment involves harvesting the crop at the start of the ripening process and cutting the culms close to the ground or uprooting the whole plants. Harvesting is preferably done in the morning so that morning dew holds the loose grains in place as much as possible. The harvested plants are allowed to after-ripen on the field proper before they are transported to the threshing floor. During the after-ripening, reallocation of water from the grain kernels to the vegetative parts takes place. In this way, the grain kernels still become sufficiently dry to be stored for a long time. If the intention is to produce *'frikeh'*, part of the crop was processed without further after-ripening.

1.1.2.4 *Modelling domestication*

Barley and wheat plants with a non-brittle rachis are naturally found in small numbers in populations of wild cereals. By atmospheric radiation, a gene that codes for the fragmentation of the rachis can mutate on a chromosome present in an ovule or a pollen grain. The grain kernel with such a mutation develops in the spike and can still be dispersed. The plant that grows out of this mutated grain kernel will produce a spike of which all grain kernels will have the same mutation. But the rachis of the mutant plant is non-brittle and the plant cannot disperse its grain. As a result, the number of mutant plants will not increase in number and the grain kernels will most probably be eaten by animals.

When a population of wild cereals is harvested by hand stripping or by using a swinging basket, only the ripe, loose, upper units of the spike will be collected; the spikes of the mutated plants, which have a tough rachis, will not be harvested. But when a population of wild cereals is harvested by uprooting or by using a harvesting knife or sickle, the spikes of the mutated specimens will be harvested as well.

Because the grain of wild plants is characterized by asynchronous *'germination'*, the ripening of the grain within a population will show some variation. As a result, culms that are gathered by hand will mostly probably also include a few culms that are not yet fully ripe. Breaking off this bunch of culms by hand is not easy; the use of a harvest knife or sickle offers a solution. Thus the use of a harvesting knife or sickle, most probably related to the optimization of the harvest, appears to have played a role in the domestication process.

Initially, the harvest of spikes with an intact rachis must have taken place unconsciously, due to the low frequency of the mutated plants in wild populations. Only when plants with an intact rachis increased in number did conscious selection become possible. Such an increase could have resulted from using a substantial part of a harvest collected by uprooting or by using a sickle for sowing into new fields (Hillman & Davies, 1992). The production of side shoots (tillering) of such mutated plants may also have contributed to their recognizability and attractiveness. By harvesting the spikes of the mutated plants separately and sowing them in new plots, a rapid shift in the direction of a fully domesticated crop could have taken place. The selective harvesting from individual plants with special characteristics prior to the regular harvest is still a proven method that ensures high-quality seeds (Hamilton, 2003).

The process of sowing is, unfortunately, not recognizable in archaeological research. Special tools that can be used during sowing (and that might be recognizable archaeologically) have only recently been developed. In the absence of such tools, sowing units can easily be carried in a fold of clothing or in any container. However, when examining subfossil threshing remains of barley and wheat, it is possible for archaeobotanists to deduce that a particular crop must have been sown by noting whether the scars on the *'rachis fragments'* are smooth or irregular. Smooth scars indicate wild grains, whereas irregular scars indicate domesticated grains, which must by necessity have been sown. From such archaeobotanical research, it has become clear that barley and wheat had become fully domesticated in south-west Asia by around 8300 BCE. Even the naked forms of barley and wheat were available by then. Sowing is necessary in the cultivation of hulled grain, but is also used in the cultivation of naked grain to optimize the harvest.

1.1.2.5 *Domestication and crop selection*

Even before pottery became available in south-west Asia, late hunter-gatherers and early farmers could choose from a broad spectrum of crops with maximum genetic variation. This offered humans the opportunity to try out which crops are best adapted to the local environmental conditions, as determined by climate and soil. The further selection of crops that are adapted to local environmental conditions could then be tested against economic criteria. These economic criteria deal with the *'yield'* (the economic value of the *'principal product'* [alternative term: *'primary product'*] and the *'by-product'* [alternative term: *'secondary product'*]); the input of energy from labour and fuel (for production, harvesting, crop processing, and food processing); and the efficiency of storage and transport. For cereals, this selection took place on the level of the subspecies (and with barley also with regard to hull tightness). It is notable in this respect that many common names exist for subspecies of barley and wheat but that common names are lacking at the species level, likely because the species level does not matter all that much in crop selection.

Crop selection is accompanied by the development of the specific technology and the material culture that is adapted to the processing and food preparation of that particular crop. The control of fire, which is necessary for heating the food, had long since been achieved (Goudsblom, 1992). Separating the grain kernels from their chaff, *'bran'*, and *'germs'*, and then further granulating or *'pulverizing'* the grain, became possible by using vessels and mortars and grinding stones (Eitam et al., 2015; Pederson et al., 2016).

FIGURE 10: Collecting vegetables on a hillside. Some of the vegetables are offered for sale in local markets (Mir Deh, Iran; April 2017; RC).

FIGURE 11: Selling vegetables that have been collected in the wild. The vegetables include wild onions (*Allium* spp.), *Chaerophyllum macrospermum*, and Water cress (*Nasturtium officinale*) (Saqqez, Iran; April 2017; RC).

The transition from hunting and gathering to agriculture is therefore to be regarded as a gradual process, with a somewhat arbitrarily defined turning point when it comes to domestication. Moreover, hunting and gathering was not abandoned during the transition to agriculture; it is a welcome addition to the annual harvest of the fields. Such additional collection of wild plants still takes place today. Successful plant gathering requires in-depth knowledge that is passed on from generation to generation (Ertuğ, 2009; figures 10 and 11).

1.1.3 Use of milk

Although the domestication of sheep and goat dates back to the 9th–8th millennium BCE and that of cattle to the late 7th millennium BCE, the first evidence for the use of milk, confirmed through residue analysis of pottery, has been dated to the late 7th millennium BCE (personal communication P.M.M.G. Akkermans, 2018; Evershed et al., 2008; Zeder, 2011). Because this date coincides with the early Pottery Neolithic (PN), the question arises whether milk was being used even earlier than the late 7th millennium, but in a manner that we are not yet able to trace archaeologically. Indeed, contemporary traditional milk processing shows that such processing can be achieved without the use of pottery.

Initially milk was only available during the period in which young animals are born and drink milk. During this period, milking by humans could be combined with suckling by the young animals. Simultaneous milking and suckling is possible because the number of nipples is two (in sheep [*Ovis aries*] and goat [*Capra aegagrus*]) or four (in cow [*Bos taurus*], yak [*Bos grunniens*], water buffalo [*Bubalus bubalis*], camel [*Camelus dromedarius*], dromedary

[*Camelus ferus*], and horse [*Equus ferus*]). It is especially feasible in larger animals, because the young animal can stand on one side of the lactating mother, while the milking is done from the other side (see, for example, Ferdinand [1993:71] with respect to the dromedary in Qatar and Schmid & Bendl [1997:71] with respect to the camel in Mongolia). It is also possible to do the milking and suckling sequentially, in which milking is done first, to obtain the milk that has the highest fat content, and the remaining milk is left for the young animal. In order to ensure that the mother's milk becomes available in the udder for milking, it may be necessary to allow the young animal to approach its mother (Edelberg & Jones, 1979).

Continuous milking will have prolonged the period during which the milk was available. The milking period of sheep is 2–3 months and that of goats is 4–5 months, but only in summer. Cows can be milked for a longer period and the yak even year-round (Jones, 1996; Palmer, 2002). The size of the flock as well as the duration of the period of milking will determine the amount of milk. The maximum size of a flock of sheep and/or goats is determined by the control and care provided by the person(s) who take(s) care of the animals. According to Edelberg and Jones (1979), a shepherd in Nuristan (north-western Afghanistan) can herd up to 100 goats or up to 20 cows. A similar number of goats is mentioned by Ferdinand (1993) for the Bedouins of Qatar: a flock of some 200 goats was looked after by two shepherds. These animals belonged to 5–6 families, meaning that, on average, each family thus owned some 30–40 animals.

The treatment of milk differs fundamentally from that of grain. Whereas grain is harvested once a year in bulk quantities, milk is obtained daily, over a considerable period of time. If harvested in a ripe state, grain has a long storage life. Only if grain is harvested while still unripe is it necessary to process the grain in order to reduce its moisture content. The large-scale conversion of grain into a (semi-)edible food with a long shelf-life is related to the efficient use of fuel and the killing of microorganisms and small animals (including insects and mites) that have infected the harvest.

The production of a daily milk surplus, on the other hand, necessitates processing the milk into a product that can be kept for some time without spoiling. This entails that milk processing needs to take place on a daily basis during the entire milking period. If the animals return home daily, milk can be processed into either a dry '*milk product*' or a dry '*cereal-milk product*' (alternative term: '*milk-cereal product*'). Even if they do not return home daily, their milk can still be processed for storage. Shepherds who take care of flocks of animals during transhumance, and whose animals are used for milk production as well, have the responsibility to convert the milk into a dry milk product. Such dry milk products can be collected by family members from a temporarily campsite, before the shepherd(s) and the animals move to another area (Edelberg and Jones, 1979).

1.2 Morphology of barley and wheat

1.2.1 Pre-harvesting

The attachment of the flowers to the stem differs among plant groups. To describe these differences in a systematic manner, researchers have devised and used a typology of the different groupings. In the flowering stage, this organization is called '*inflorescence*', and in the fruiting stage, it is called '*infructescence*'. Because fruits develop within flowers, the organization of flowers and fruits is the same in a given species and the name used to designate the inflorescence can also be used to designate the infructescence.

The infructescence of both barley and wheat is a spike (figure 12). The spike consists of a central axis (rachis) that can be considered an extension of the culm (stem). Both the culm and the rachis have a number of thickenings. These thickenings are called '*nodes*', and the parts between the nodes are called '*internodes*' (figure 13; here shown after threshing). The leaves are attached to the culm nodes, whereas the spikelets are attached to the '*rachis nodes*'.

FIGURE 12: Spike and upper part of culm of domesticated hulled 2-row Barley (*Hordeum vulgare* ssp. *distichon*) (left) and domesticated Emmer (*Triticum turgidum* ssp. *dicoccon*) (right) (DF).

A spikelet is the basic unit from which the spike is built. A spikelet consists of a '*rachilla*', one or more flowers, and hulls. After fertilization, the fertile flowers each produce a grain kernel. The hulls consist of reduced leaves whose origin is comparable to the leaves attached to the culm. The hulls are green in the flowering stage and the early ripening stage, during which period photosynthesis takes place. Three types of hulls are recognized: the '*glume*', the lemma, and the palea. Glumes are spiny in barley and scaly in wheat, whereas the lemma and palea are scaly in both genera. The top of the glume and the lemma can narrow into an awn. The hulls protect the miniscule flowers and the grain kernels.

The combination of the lemma and palea with the flower or grain kernel is called '*floret*'. The combination of two glumes and one or more florets is called '*spikelet*'. There are structural differences between the spike in barley and the

FIGURE 13: Intact spike and upper part of culm (left) and upper part of the spike after threshing (right) of domesticated hulled 2-row Barley (*Hordeum vulgare* ssp. *distichon*). The intact spike is shown in side view, and the spike after threshing is shown in ventral–dorsal view. In the latter, the central fertile florets have become detached from the rachis, whereas most of the glumes of the central spikelet, as well as the sterile lateral spikelets, are still attached to the rachis (DF).

spike in wheat, and these differences have consequences for both crop selection and crop processing.

The rachilla is the main axis of the spikelet. In barley, the single floret is attached to the rachis node, and the rachilla does not support another floret. If present, the rachilla is visible at the basal part of the ventral groove (clearly visible in figure 21). In wheat, the lowest floret is attached to the rachis node, and the subsequent florets of the spikelet are attached to the much reduced rachilla.

In barley, three spikelets are attached to a single rachis node. Each spikelet contains only one floret, which means that in the fruiting phase each fertile spikelet contains only one grain kernel. The glumes of each spikelet are needle-shaped and only slightly widened at the base. Although the spiny glumes do not enclose the flowers and grain kernels, they are important in seed dispersal.

In wild Barley (ssp. *spontaneum*), and also in the first domesticated 2-row Barley (ssp. *distichon*), only the middle spikelet is fertile. The lemma of this central, fertile spikelet is characterized by a long, sturdy awn. Both lateral spikelets are sterile and often have short stalks, and the lemma does not have a long awn. A cross section of the spike of both wild and 2-row barley shows the central spikelets containing the grain kernels to be clearly arranged in two rows, giving the spike a flat appearance. In domesticated 6-row barley, both lateral spikelets are also fertile. A cross section of the spike of 6-row barley shows the central and lateral spikelets arranged in six rows.

In wheat, a single spikelet is attached to each rachis node. In wild Einkorn (*Triticum monococcum* ssp. *aegilopoides*) and wild Emmer (*Triticum turgidum* ssp. *dicoccoides*), a spikelet consists of two glumes and three florets. In einkorn, usually only one of the three florets produces a grain kernel. In emmer, usually two florets produce a grain kernel. However, abnormalities in fertility and grain production do occur. For example, there are instances of einkorn with two grain kernels in a spikelet and emmer with a grain kernel in all three florets (Cappers & Neef, 2012; Willcox, 2005).

1.2.2 Harvesting and post-harvesting

The names available for the description of the individual plant parts are usually unambiguous. But it is remarkable that, when it comes to seed dispersal, harvesting, and post-harvesting processing, the respective plant units often have no unambiguous names. Here, we present names that are in accordance with the botanical description and that also support the understanding of the specific processes applied to the different crops in order to convert them into units that can be used in food preparation (figure 14).

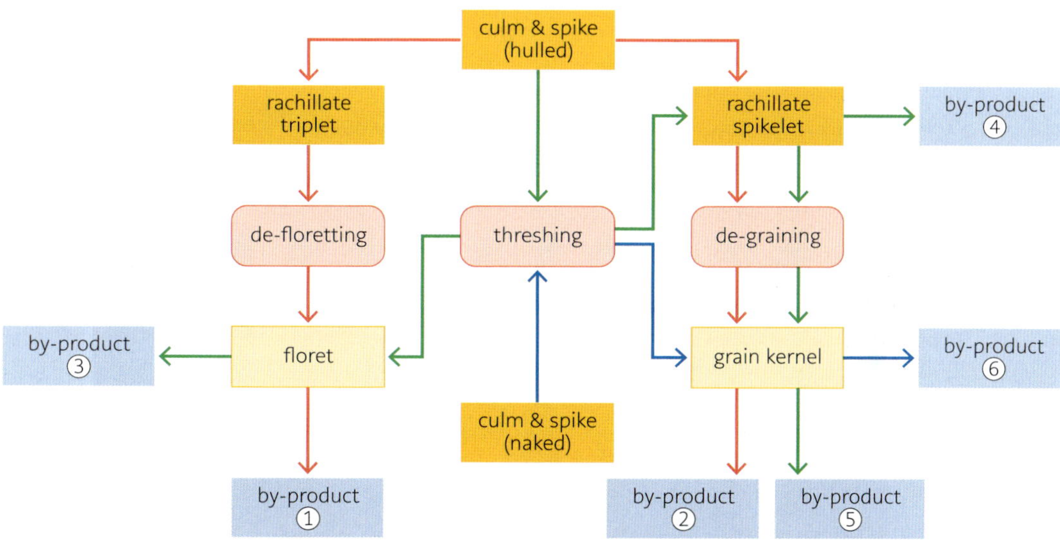

FIGURE 14: Harvesting units of barley and wheat (shown in bright yellow) and their fragmentation into post-harvesting units. The post-harvesting units are classified into edible principal products (shown in light yellow) and their separated non-edible by-products (shown in grey; see figure 37 for a description of their composition). Three fragmentation processes are presented: de-floretting, de-graining, and threshing. The combination of the ripeness of the grain kernel, rachis brittleness, and hull tightness is indicated with coloured arrows (⟶ = harvesting ripe units from wild cereals; ⟶ = harvesting partly unripe spikes from wild cereals and harvesting fully ripe spikes from domesticated hulled cereals; ⟶ = harvesting partly unripe or fully ripe spikes from naked cereals).

The harvesting unit in both barley and wheat can be a diaspore, a spike attached to a part of the culm, or the uprooted whole plant with root fragments attached to the basal part of the culm. Because the focus of this section is on the fragmentation of the infructescence, the description 'culm & spike' is used in figure 14 irrespective of the length of the culm and the presence of root fragments.

A diaspore is the natural dispersal unit of a plant, and such a unit can be harvested from both wild barley and wild wheat. Usually, the term 'spikelet' is used instead, which is a *pars pro toto* for a combination of a spikelet and its supporting rachis segment. Terms for the diaspores of wild barley and wild wheat that take into account both parts of the dispersal unit are *'rachillate triplet'* for barley (figures 15 and 16) and *'rachillate spikelet'* for wheat (figures 17, 18 and 19; Cappers et al., 2016).

A rachillate triplet consists of a rachis segment and three spikelets, whereas a rachillate spikelet consists of a rachis segment and a single spikelet. The rachillate spikelet can be obtained from both wild wheat and domesticated wheat. Ripe rachillate spikelets from wild wheat are characterized by smooth scars marking the former attachment (figure 17), whereas those from domesticated, hulled wheat are characterized by irregular scars (figure 19).

The edible parts that are obtained through post-harvest processing are the *'floret'* and the *'grain kernel'*. The floret is the typical unit of hulled barley; it can be obtained from rachillate triplets (harvested from wild barley) or from whole spikes (harvested from domesticated barley). In both cases, the floret has to be separated from the rachis. This fragmentation process is called

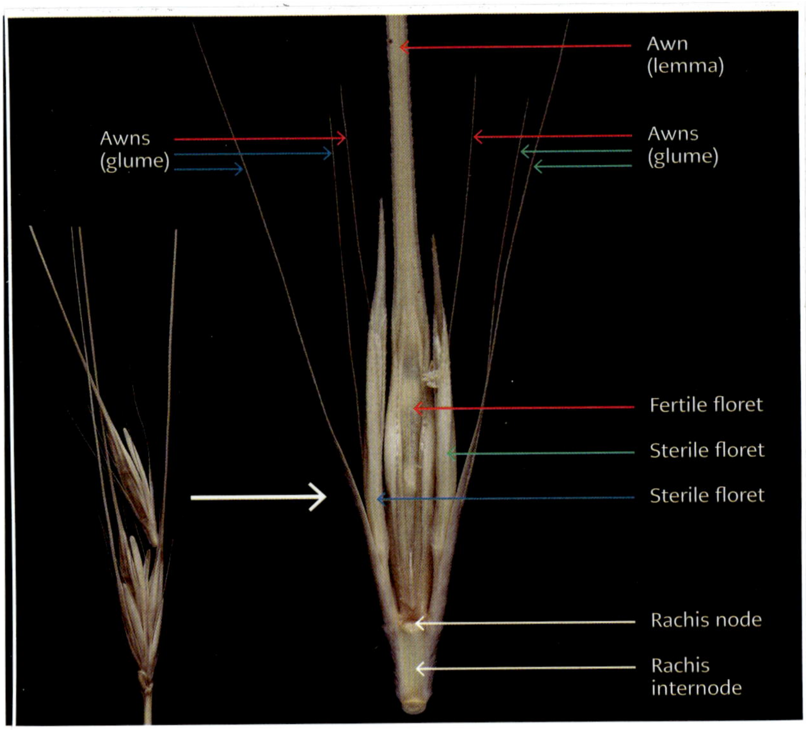

FIGURE 15: Disarticulating spike (left) and rachillate triplet (right) of wild Barley (*Hordeum vulgare* ssp. *spontaneum*). Only the lower parts of the awns of the lemma of the central spikelets are shown. The glumes are small and gradually narrow into a long awn. In this and the following figures, the units of the three individual spikelets are each labelled with their own colour (DF).

'de-floretting' when applied to the rachillate triplet (figure 16) and *'threshing'* when applied to whole spikes (figure 21). De-floretting and threshing are followed by winnowing and/or sieving to separate the florets from the vegetative plant parts.

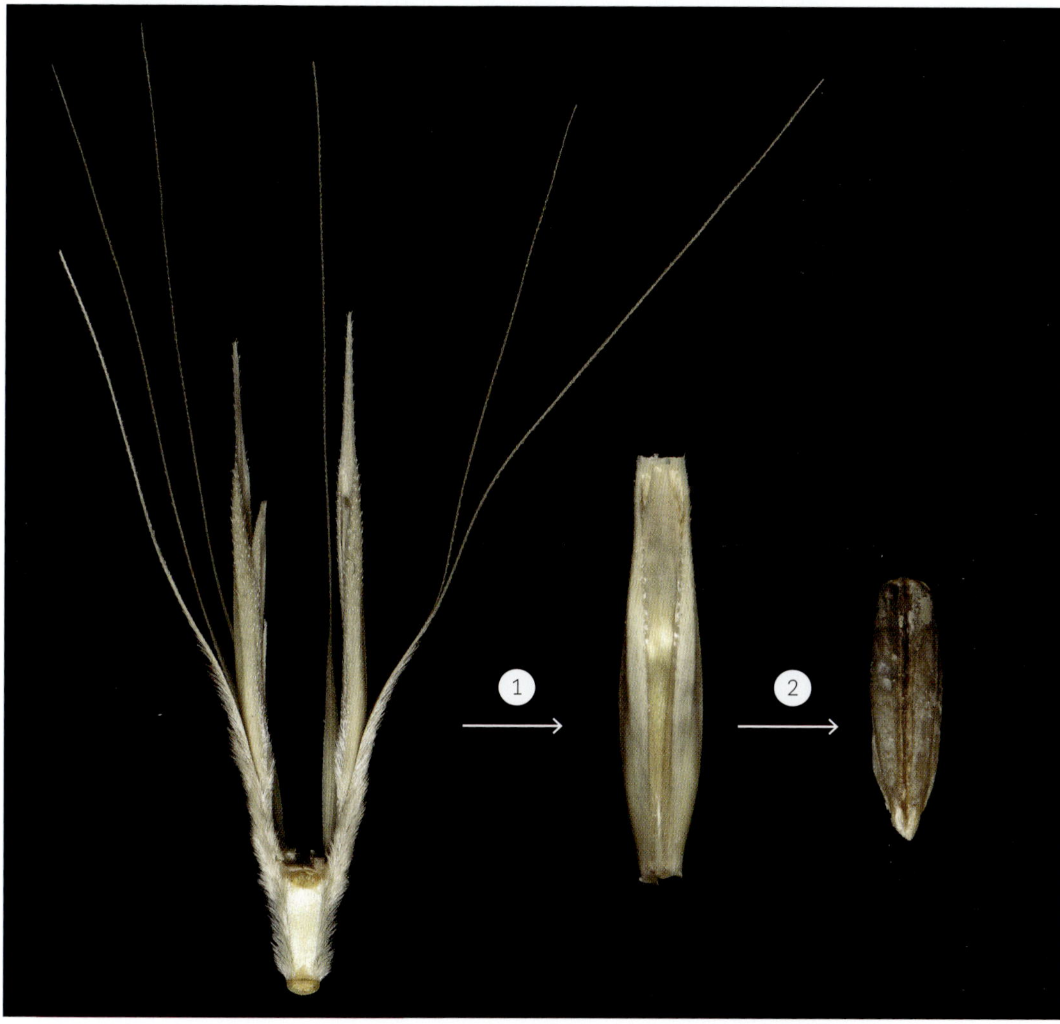

FIGURE 16: Fragmentation of a rachillate triplet of wild Barley (*Hordeum vulgare* ssp. *spontaneum*): (1) de-floretting and (2) de-graining. Both kinds of fragmentation have been done manually. De-graining has been performed to show the proportion of grain kernel to chaff (DF).

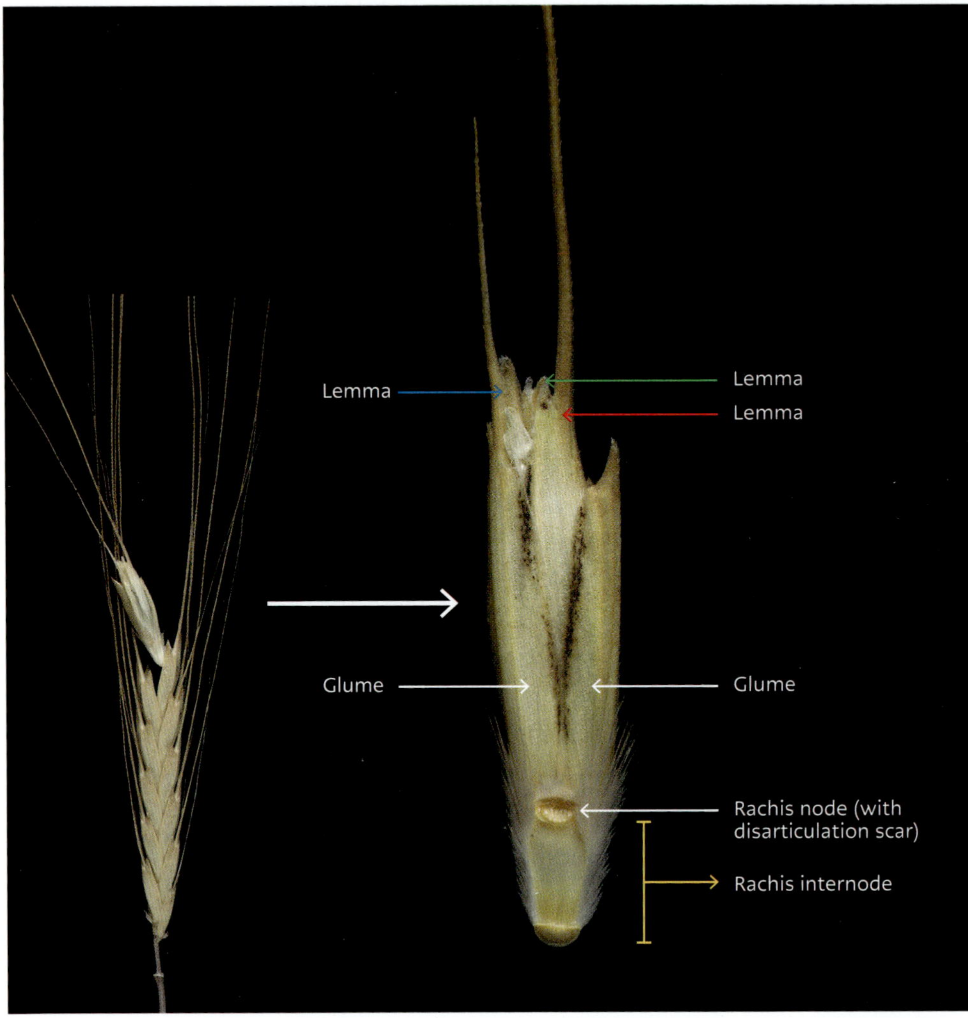

FIGURE 17: Disarticulating spike (left) and rachillate spikelet (right) of wild Einkorn (*Triticum monococcum ssp. aegilopoides*). Only the lower parts of the awns of the lemma of the central spikelets are shown. Both glumes have a tooth, and the first and second lemma are awned. The disarticulation scar at the rachis node is regular, and the '*rachis internode*' is complete and has a smooth scar at its base (DF).

FIGURE 18: Fragmentation (de-graining) of a rachillate spikelet of wild Einkorn (*Triticum monococcum* ssp. *aegilopoides*). The fragmentation has been done manually to show the proportion of grain kernel to chaff (DF).

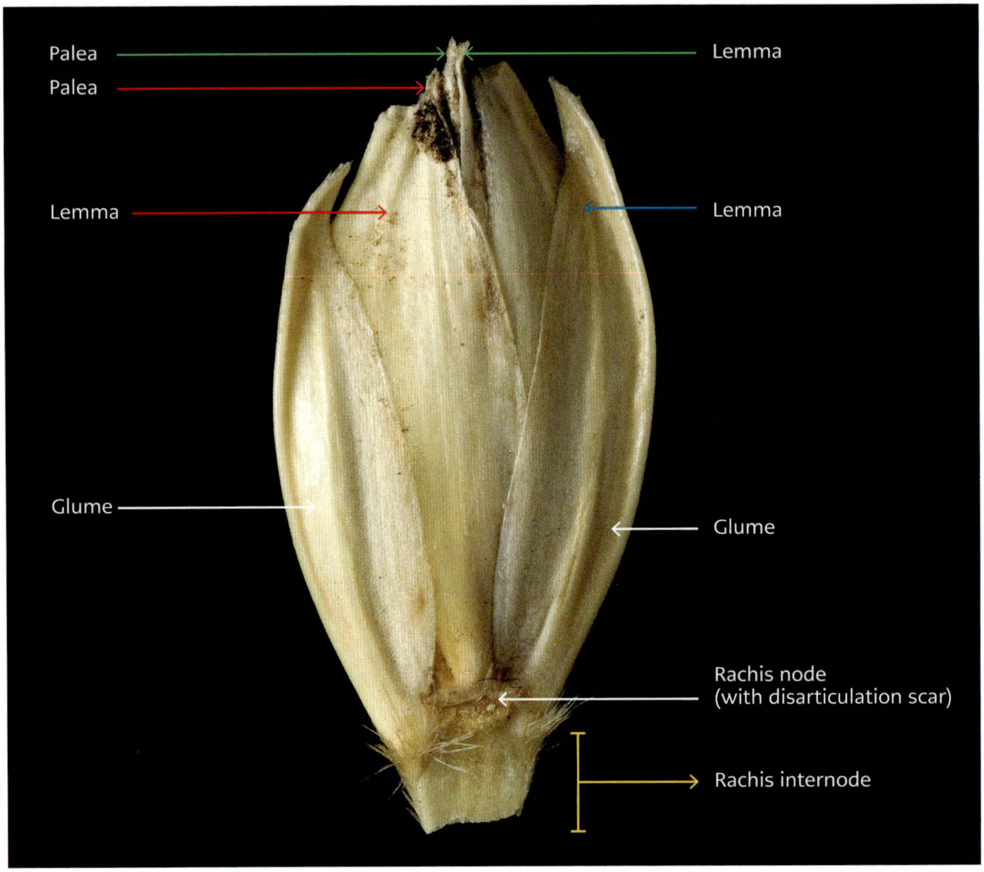

FIGURE 19: Rachillate spikelet of domesticated Emmer (*Triticum turgidum* ssp. *dicoccon*). This threshing unit consists of two broad glumes and three florets. The top is visible in only two of the three paleas. The disarticulation scar at the rachis node is irregular, and the lower part of the rachis internode has broken off (DF).

FIGURE 20: Rachillate spikelet of domesticated Emmer (*Triticum turgidum* ssp. *dicoccon*). Complete specimen obtained after threshing (bottom) and fragmented specimen showing the two grain kernels and the relatively large proportion of vegetative plant parts (top). The fragmentation has been done manually (DF).

FIGURE 21: Upper part of the rachis of domesticated hulled 2-row Barley (*Hordeum vulgare* ssp. *distichon*) after threshing. Most of the glumes of the central spikelets and the sterile lateral spikelets, each consisting of two glumes and a floret, are still attached to the rachis. A single floret is shown to scale in ventral view together with the rachis, and enlarged in ventral and dorsal view (DF).

The fruit of grasses is dry and contains a single seed. The *'fruit wall'* (*'pericarp'*) is closely adherent to the *'seedcoat'* (*'testa'*). This fruit type is called a *'caryopsis'*. The fruit of gathered wild grasses and domesticated grasses is often termed the *'grain kernel'*.

The fragmentation process that is applied to separate the grain kernel from its vegetative plant parts differs depending on the harvesting unit and the hull tightness. Removing only the lemma and palea of a floret to obtain the whole grain kernel is a delicate process that requires sophisticated technology. If the hulls have to be removed from the florets, its removal usually also results in the removal of the fruit wall, seedcoat, *'aleurone layer'* and germ. This combined removal is called *'pearling'* and is typically used in the processing of hulled barley (figure 22; see also section 2.2.2).

FIGURE 22: Pearled florets of hulled Barley (*Hordeum vulgare*). Remains of the hulls (palea) are still present in the hilum groove (Tabriz, Iran; June 2017; DF).

In hulled wheat, the lemma and palea do not adhere to the grain kernel. This makes it possible to isolate the whole grain kernel from the vegetative plant parts of the rachillate spikelet, although a separate fragmentation process is required to achieve this. This fragmentation process is termed *'de-graining'* (alternative term: *de-hulling*; figures 18 and 20). In naked barley and wheat, the grain kernel is obtained when the whole spike is fragmented through threshing (figures 23 and 24).

Characteristic features on the outside of the grain kernel are the hilum groove at the ventral side (*'adaxial'*: facing towards the rachis) and the embryo area at the dorsal side (*'abaxial'*: facing away from the rachis). In barley, the hilum groove is shallow (figure 23), whereas in wheat the hilum groove is rather deep (figure 24). This difference has an impact on the way the grain kernel becomes fragmented during granulation (see figure 31).

The seed consists of the seedcoat, the aleurone layer, the *'starchy endosperm'*, and the *'embryo'*. Although the aleurone layer is part of the *'endosperm'*, it has a discrete tissue and separates from the starchy endosperm during polishing and pearling (see section 2.2.2).

FIGURE 23: Empty spike and grain kernel of domesticated naked 2-row Barley (*Hordeum vulgare* ssp. *distichon*). A single grain kernel is shown to scale in ventral view together with the empty spike, and enlarged in ventral and dorsal view (DF).

FIGURE 24: Empty spike and grain kernel of Hard wheat (*Triticum turgidum* ssp. *durum*). A single grain kernel is shown to scale in ventral view together with the empty spike, and enlarged in ventral and dorsal view (DF).

1.3 Nutritional value of barley, wheat, and milk

The four basic food molecules are water, carbohydrates, proteins, and lipids. Water (H_2O) consists of two hydrogen atoms (H) and a single oxygen atom (O). Due to its electrical polarity, it is a solvent for compounds that also have some electrical polarity, such as carbohydrates and proteins. Daily intake of water is essential in human nutrition.

Carbohydrates consist of carbon (C), hydrogen (H), and oxygen (O), and their general formula is $C_x(H_2O)_y$. Carbohydrates exist in many different molecules. Their main function in plants is to store energy that is fixed from solar radiation and to provide supporting structure in the plant tissues. Simple carbohydrates include molecules with five carbon atoms, which are the main components of DNA and RNA, and those with six carbon atoms, which include glucose. Glucose is the molecule from which the stored energy can be released by splitting the molecule into smaller units. Glucose is soluble in water and can be transported to each part of the organism. Starch is a large-chained carbohydrate that is insoluble in water and therefore does not affect the water activity. It contains much energy, which becomes available once it has been split into glucose and water.

Proteins are a group of molecules that provide structure and are essential in catalyzing biochemical processes. The building units of proteins are some 20 amino acids. Amino acids contain basic amino (NH_2) and acidic caroboxyl (COOH) groups. Their number and sequence within a particular molecule determine the primary structure of each protein. The nutritional value of proteins in food is determined by the presence of the so-called essential amino acids in particular because these amino acids cannot be synthesized from digested food but have to be present in the food as amino acids.

Lipids are a heterogeneous group of compounds that include fats and oils. They lack electric polarity and are not soluble in water. Lipids are suitable for providing structures, such as cell membranes, but are also used for the efficient storage of energy.

1.3.1 Barley and wheat

Grain kernels consist of water, carbohydrates, proteins, lipids, vitamins, minerals, and fibres. These components are unequally distributed within the grain kernel. Therefore, the nutritional value will depend on the kind of unit that is used in food production. In barley, the unit can be a floret, and in both barley and wheat, it can be either the whole grain kernel or the grain kernel from which the bran and germ have been removed.

The water content of ripe grain kernels is low, and this makes it possible for grain to survive in the soil for some years or, similarly, to be stored for some years after harvesting. Grain kernels are particularly rich in carbohydrates, which are mainly present as starch and are stored in the starchy endosperm. This entails that proteins and lipids are present in only minor quantities. In barley, the outer layers of the grain kernel are rich in proteins, whereas essential amino acids and lipids are concentrated in the aleurone layer and the embryo; the water-insoluble, gluten-forming storage proteins are concentrated in the starchy endosperm (Briggs, 1978). In wheat, in grain kernels that are relatively small, such as those of the Einkorn (*Triticum monococcum* ssp. *monococcum*), the proportion of proteins is relatively large in comparison with wheat taxa with larger grain kernels (Hildalgo & Brandolini, 2013).

Dietary fibre is present in the hulls, the fruit wall, and the seedcoat. Therefore, in barley, the food will contain this fibre if either florets (from hulled barley) or unpolished grain (from naked barley) are used in food processing, whereas in wheat, the food will only contain this fibre if unpolished grain is used.

1.3.2 Milk

'*Raw milk*' (alternative term: '*fresh milk*') consists of water, carbohydrates, fat, proteins, and minerals. The relative proportion of each of these nutrients depends on the kind of animal and on the plant foods that it has eaten. For example, sheep's milk contains less water and up to twice as much fat and protein than goat's or cow's milk. Milk processed in factories today is mixed to obtain a standard composition. The average proportion of water, carbohydrates, fat, and proteins of such cow's milk is presented in bar 5 of figure 25.

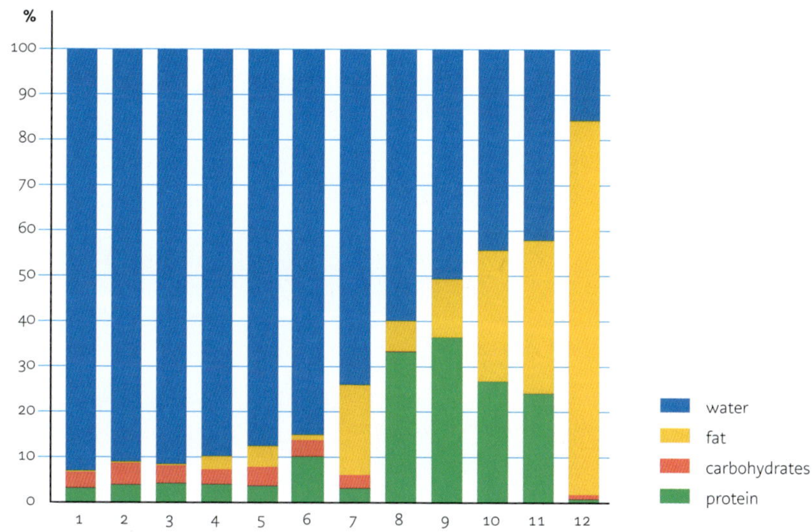

FIGURE 25: Composition of raw cow's milk and related milk products per 100 grams. The milk products are sorted by decreasing moisture content. 1: buttermilk; 2: skimmed milk; 3: low fat yoghurt; 4: full fat yoghurt; 5: raw milk; 6: fresh buttermilk curds; 7: sour cream; 8 cheese (10+); 9: cheese (20+); 10: cheese (45+); 11: cheese (50+); 12: butter (source: NEVO-online version 2016/5.0, RIVM, Bilthoven).

The proportions of nutrients change when raw milk is converted into different milk products (alternative term: '*dairy products*'). Fat is concentrated in '*cream*' and '*butter*', and the by-products of these milk products thus have a reduced fat content (figure 25: 1–3). Fat and proteins are typically concentrated in '*cheese*' (figure 25: 8–13).

The predominant carbohydrate in milk is milk sugar (lactose), which easily becomes converted into milk acid (lactic acid) due to the omnipresent lactic acid bacteria. This changes the taste of milk from sweet into sour. Milk sugar and lactic acid are soluble in water and can be extracted from milk that is to be used for making butter and cheese.

Milk contains both vitamins that are soluble in water (vitamins B-complex and C) and vitamins that are soluble in milk fat (vitamins A, D, E, and K). Milk contains several minerals, including calcium (Ca), potassium (K), phosphorus (P), sodium (Na), and magnesium (Mg).

2
Processing of cereals and milk

2.1 Aims

It could be argued that hunter-gatherers living in south-west Asia were unlucky in having wild barley and wild wheat as the progenitors of domesticated cereals. Being wild grasses, both species produce diaspores with a relatively large proportion of inedible vegetative plant parts. And once domesticated, they have to be harvested as a spike with at least part of the culm. Because barley and wheat plants are relatively short, and therefore have to be cut relatively close to the ground, the harvest of domesticated barley and wheat usually also includes a range of arable weeds (figure 26). To obtain the edible parts, the harvest has to be threshed. For practical reasons, the bulky harvest is threshed on the ground. Although the threshing floor can be cleaned before threshing, some degree of mixing with soil, stones, and other impurities cannot always be avoided. In order to prevent clay particles becoming mixed with *'meal'* (alternative term: *'whole flour'*) or *'flour'*, they have to be removed before pounding or grinding. Stones are particularly problematic because they can cause tooth damage. And some of the diaspores of arable weeds are undesirable because they are hard (e.g. Beet, *Beta vulgaris*), spiny (e.g. wild Oats, *Avena* spp.), or poisonous (e.g. Corn cockle, *Agrostemma*

FIGURE 26: A bundle of harvested plants of Bread wheat (*Triticum aestivum* ssp. *aestivum*). The naked wheat is harvested close to the ground to facilitate after-ripening on the field proper. The reaped plants are tied together with twine. Long arable weeds, including Beet (*Beta vulgaris*), are reaped together with the crop, and their diaspores will become mixed with the grain kernels on the threshing floor (Ezbet Basili, Egypt; April 2002; RC).

githago, and Darnel, *Lolium temulentum*). The weight, size, and shape of clay particles, stones, and diaspores of arable weeds may be similar to the threshing units of the crop. In fact, cleaning by winnowing and sieving over a prolonged period has resulted in an unintended selection of diaspores of arable weeds that match the crop's threshing unit in weight, size, and shape.

A different picture is presented by, for example, the sub-Sahara. The founder crops of this region include Pearl millet (*Pennisetum glaucum*) and Sorghum (*Sorghum bicolor*). Their infructescences consist of dense panicles that ripen on long, thick culms. This enables the harvest of individual infructescences without any arable weeds or soil particles (figure 27). The grain kernels of both crops are exposed on the outside and easily threshed out by pounding in a vessel. And the grain kernels only have to be separated from the vegetative parts of the spike, not from contaminants.

FIGURE 27: A pile of harvested infructescences of Pearl millet (*Pennisetum glaucum*). The spikes are produced on long culms and are cut individually, without any arable weeds. The culms of the crop are used as '*straw*' matting to prevent contact with the soil (Taleg, Senegal; October 2016; RC).

The processing of barley and wheat is related to improved edibility and digestibility. In addition, special treatments can be applied to improve storage life and sowing. The edibility of barley and wheat is improved by fragmentation and separation processes. A final fragmentation process is either pearling or polishing, which is initially aimed at enhancing the shelf-life of the prepared food. The storage of grain, whether or not covered by hulls, is not problematic because the moisture content of the grain kernels is strongly reduced during the ripening process. Grain can therefore be stored for several years without spoiling. Par-heating can be applied to the rachillate spikelets of wheat to facilitate to removal of the chaff (see frikeh and '*bulgur*'). An advantage of this treatment is that heating also kills microorganisms and the different stadia of insects and mites present in the crop. During food processing, heating is no longer optional; heat has to be applied to the grain to improve its digestibility.

Milk processing is related to the transformation into a milk product with a long shelf-life. This is necessary when a surplus of milk is produced and cold storage is not available because milk will soon spoil if it is not kept cool. Even if it is kept cool, milk will eventually spoil unless it has been preserved by heating, by reducing its moisture content, by salting, or by increasing its acidity. Raw milk is usually heated to kill the microorganisms that are present in the milk. The drinking of boiled milk is considered to be healthy, as noted in the example of a Bedouin of the Khushmaan clan (northern part of the Eastern Desert, Egypt) who considers the neighbouring Ababda Bedouins (southern part of the Eastern Desert, Egypt) as unsound because 'Ababda eat meat raw and drink unboiled goat's milk' (Hobbs, 1989).

The inoculation of raw milk with a starter after heating, in order to improve lactic-acid fermentation, can be considered a secondary development. Because lactic-acid fermentation results in the conversion of milk sugar (lactose) into milk acid (lactic acid), a side effect of processing raw milk into '*sour milk*' and cheese is improved digestibility for people who suffer from lactose intolerance.

2.2 Basic principles

2.2.1 Classification

A single stage of crop processing and food processing can be described using different terms, relating either to the particular plant part that is being manipulated or to the specific operation that is being applied. With terms related to the particular plant part that is being manipulated, two perspectives are possible, resulting in yet more terms. For example, the fragmentation that is aimed at isolating the grain kernel from its chaff (hulls and rachis) can be designated as '*de-hulling*' (a term that is also shortened in the literature to '*hulling*') or '*de-graining*'. The same process can further be described by the specific operation that is applied, such as '*beating*', '*rubbing*', or '*trampling*', in which a link with the tool can be evident in the term that is used. Although such terms are function well in daily use, in the context of describing food processing, they may obscure the precise change that is achieved. Processes that are used in the treatment of cereals and milk deal with the manipulation of moisture content, temperature, size, and nutrients. The value of each of these variables can be increased or decreased, whereas nutrients can also change in its composition. Processes that can be applied to achieve these changes include fragmentation, separation, mixing and shaping, heating, cooling, germination, and '*fermentation*' (figure 28).

		Fragmentation	Separation	Mixing & shaping	Heating	Cooling	Germination	Fermentation
Moisture content	*increasing*			○				
	reducing		○					
Temperature	*increasing*				○			
	reducing					○		
Size	*increasing*			○				
	reducing	○						
Nutrients	*increasing*			○	○			
	reducing	○	○					
	composition						○	○

FIGURE 28: Classification of the basic processes that are used in crop processing and food processing.

Some processes are clearly recognizable but are considered as being subordinate to the main processes that are mentioned in figure 28. Examples are wetting and '*drying*', which deal with the moisture content. Wetting is considered here to be a form of mixing with water, whereas drying is considered a method of separation (see sections 2.2.2 and 2.2.3). Some processes are quite specific and at the same time are difficult to categorize, and these processes occupy a special position. An example is the '*kneading*' of butter, which is aimed at effecting the clumping of the small fat particles, and the kneading of '*dough*', which is aimed at the creation of a gluten network.

2.2.2 Fragmentation

Fragmentation that is applied to barley and wheat differs with respect to both the aim of the process and the technology that is used. Fragmentation can be aimed at separating two components or reducing their size. Fragmentation processes that are followed by separation of the principal product and the by-product include threshing, de-floretting, de-graining, '*de-awning*', '*de-branning*', '*de-germing*', polishing, and pearling. Granulation and pulverization are fragmentation processes that are not followed by separation (figure 29). If two components are obtained, the fragmentation is aimed at improving the edibility, digestibility, and/or shelf-life. In cereals, this is valid for the principal product, whereas in milk this is valid for both components. If fragmentation is not followed by separation, it is related to mixing with other ingredients and efficient use of fuel to obtain small, uniform particles. The technology of fragmentation can be differentiated according to the direction of the movement.

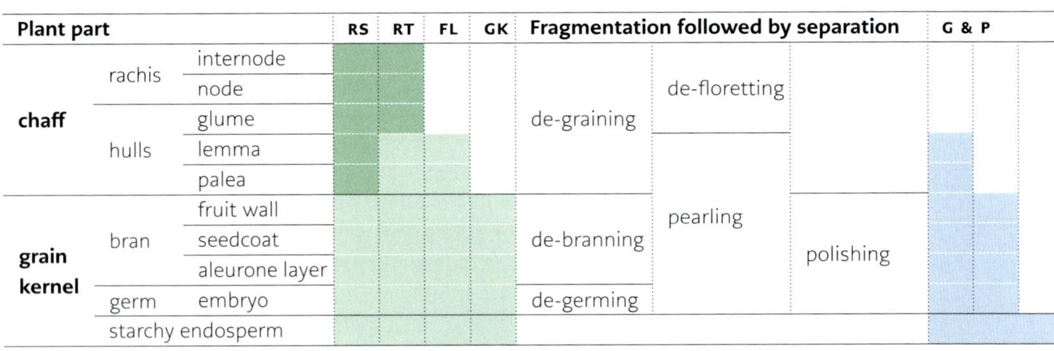

FIGURE 29: Composition of rachillate spikelet (RS; wild and domesticated wheat), the rachillate triplet (RT; wild barley), floret (FL; wild and domesticated barley), and grain kernel (GK; barley and wheat) and related processes of fragmentation. Plant parts that are non-edible are indicated with dark green; those that are edible, with light green. Granulation and pulverization (G&P) can be applied to the floret, the grain kernel, and the starchy endosperm. The plant parts concerned are indicated with blue. The fragmentation of awns (de-awning) is not indicated.

Fragmentation that is followed by separation results in a principal product and a by-product. In the treatment of domesticated barley and wheat, the first stage of this kind of fragmentation is threshing (alterative spelling: '*thrashing*'), which is aimed at removing the vegetative parts of the harvest that are easily separated. The principal product of threshing is suitable for storage. The second stage of this kind of fragmentation is aimed at removing any vegetative parts that are still connected to the grain kernel.

The term commonly used for this second stage of fragmentation is *'de-hulling'* (see also the discussion on *'second threshing'* in the Glossary). But given that the removal of only the hulls does not occur in the processing of either hulled barley or hulled wheat, and given the different compositions of the principal products that are obtained in hulled barley and hulled wheat, it makes sense to use different terms for the fragmentation of their respective harvesting and threshing units (viz.: *'de-floretting'* and *'de-graining'*; see section 2.3.3).

The third and last stage of fragmentation, which is followed by separation, deals with the removal of the bran and the germ. The removal of the bran is called *'de-branning'*, and the removal of the germ is called *'de-germing'*. Removing bran and germ from the grain kernel by individual processes is difficult, and traditional crop processing therefore usually combines de-branning and de-germing in a process called *'polishing'* (alternative but not preferred term: *'whitening'*). Due to the difference in hull tightness between barley and wheat, the second stage (de-floretting) and the third stage (polishing) are usually combined in the processing of hulled barley. This combined crop processing is called *'pearling'*.

Removal of bran and germ from the grain kernel facilitates drying of the grain without it getting mouldy. This kind of fragmentation makes sense if the intention is to use the grain to make cereal products that will be sun-dried. Polishing and pearling are applied to moistened grain and florets, and the mixture that is thus obtained has to be dried before it can be separated. Polishing and pearling is either done on a limited scale, dealing only with the quantity that is to be made into a food product, or on a larger scale, prior to storage (figure 30). If par-boiling is applied before polishing or pearling,

FIGURE 30: Polished grain kernels of Hard wheat (*Triticum turgidum* ssp. *durum*) have been sun-dried on the roof of a farmhouse. The polished grain is separated from the light bran and germ particles by winnowing. The bran and the germ are not collected; they blow away and will contribute to the settlement noise (Demirci, Turkey; September 2000; RC).

nutrients from the bran are reallocated into the starchy endosperm and the nutritional value is not much reduced. If par-boiling is not part of the crop processing, it is healthy to keep the bran and germ within the food web, either by using it in another *'dish'* or by using it as fodder for animals whose milk, eggs, or meat are eventually eaten. In Iran, the bran is mixed with *'yoghurt'* (personal communication, T. Talebi, 2017).

Fragmentation of the grain kernel is not aimed at further separation; rather, it is aimed at reducing the time that is needed for heating and thus reducing the amount of fuel required. Fragmented grain is also suitable for mixing with other ingredients. Coarse fragments are obtained by granulation, and fine, powdery fragments are obtained by pulverization, although a mixture of different fragment sizes can be obtained during the first stage of this fragmentation. Rather than continuing fragmentation until the entire quantity of grain is pulverized, it also possible to separate the mixture by sieving into several size classes (grading), in which each size class is used for a specific dish. Pulverization of unpolished grain results in meal, whereas pulverization of polished grain produces flour. Pulverizing of florets of barley produces meal that is enriched with small particles of the hulls (lemma and palea). Here, separation can be applied in a final stage (see next section).

Fragmentation is traditionally done by either vertical movement or horizontal movement. The vertical movement is termed pounding. The equipment used for pounding consists of a fixed concave mortar and a mobile, convex-ended pestle. Both can be made from stone or from a softer material, such as wood. The horizontal movement is termed grinding, which is done with a saddle quern or a rotary quern. A saddle quern consists of two unequally shaped stones: an elongated, fixed lower stone, with a flat or somewhat concave upper face, and a bolster-shaped, mobile upper stone. A rotary quern consists of an upper mobile stone, called the handstone, that rotates over a lower stationary stone. Experimental research has shown that fragmentation that is aimed at separating the principal product from the by-product is most efficiently done by pounding and that the production of meal and flour is most efficiently done by grinding (e.g. Meurers-Balke & Lüning, 1990; Samuel, 2000; Tzarfati et al., 2013).

Grinding and *'milling'* are sometimes considered as a more general concept that includes both grinding and pounding (e.g. Kraybill, 1977; Shoemaker et al., 2017). Here, the term milling is restricted to grinding in connection with a mill, which is a stage of further specialization.

The barley grain kernel has a shallow and rather broad hilum groove on the ventral side, which is present over its entire length. The barley grain kernel is therefore easily broken crosswise when granulated. The wheat kernel, on the other hand, is characterized by a rather small and deeply furrowed hilum groove on the ventral side, which is also present over its entire length. Therefore, the wheat grain kernel easily fragments lengthwise when granulated. Further fragmentation occurs crosswise and is more irregular (see figure 31).

A special form of fragmentation is washing. It can be applied to cereals and butter. *'Ordinary butter'* is clarified by washing out any remaining proteins and milk sugar to improve its shelf-life. The washing of florets and grain is aimed at removing impurities or certain dissolved nutrients. Florets and grain kernels may be washed to remove clay particles, dirt, or even urine and faeces

FIGURE 31: Fragments of barley and wheat showing the size of the fracture surfaces. In the fragments that have only been fragmented lengthwise, the width of the fracture surface is indicated with an arrow. (1) pearled 2-row Barley (*Hordeum vulgare* ssp. *distichon*) from Morocco; (2) bulgur of Einkorn (*Triticum monococcum* ssp. *monococcum*) from Turkey; (3) bulgur of Hard wheat (*Triticum turgidum* ssp. *durum*) from Turkey; (4) bulgur of Bread wheat (*Triticum aestivum* ssp. *aestivum*) from Turkey. Most probably fragmentation was achieved through grinding (DF).

that became mixed with the grain during threshing with the use of animal power. Washing can also be applied to remove the excess of salt that may have become mixed with the grain during storage. Halstead (2014) mentions that in

Metaxada (southern Greece), 'anyone who had time washed the grain to produce whiter bread', suggesting that people give a cultural rather than an economic argument for this process. Halstead also mentions that washing was applied to grain infected with rust.

Clay particles are a regular component of the impurities present in stored grain (Cappers et al., 2016). The clay particles originate especially from the roots that are harvested during uprooting or during cutting with a sickle. In the latter case, some culms will be uprooted due to the operator pulling at the culms to obtain the desired tension prior to cutting them. Another source of clay particles is uprooted tall plants that are collected prior to harvesting with a sickle and that are used for tying the plants into bunches. Washing the grain is preferably done prior to granulation and pulverization because at that point in time the size difference between the dissolved clay particles and the florets or grain kernels is at a maximum, making sieving very effective. Small quantities of grain can be processed by wet sieving, whereas large quantities can be washed and then sieved after drying.

2.2.3 Separation

Separation is a process that is applied to both cereals and milk. In cereals, separation primarily concerns the solid components, whereas in milk separation takes place within the liquid. Separation of water from grain usually takes place during the final ripening stage of the standing crop. If the crop is harvested unripe, excess water is removed by heating (resulting in the production of frikeh; this can be applied to both hulled and naked cereals) or by reallocation of water from the grain into the vegetative plant parts during after-ripening (in the case of naked cereals). The separation of solids is based on differences in weight, size, and shape. Separation by weight is done by winnowing, in which the mixture is allowed to fall downwards while wind is acting on it. If combined with sieving, winnowing is always the first step in separation.

Winnowing can involve a separate implement, although sieves are also used for both winnowing and sieving. Although non-sieve winnowing can be done with simple implements, such as a flat fan, this method requires the presence of a breeze that is neither too strong nor too weak.

Sieving can be done with a single sieve or with a combination of a coarse sieve and a fine sieve. If a single sieve is used, the size of the meshes determines if the principal product is retained on the sieve or passes through it. If florets of barley are pulverized, it is possible to separate coarse fragments of the lemma and palea with a meal sieve. Such a sieve has a screen with a very small mesh aperture; it can be made, for example, by weaving long hairs. The meal passes through the screen, and the coarse fragments of the hull remain as impurities on the screen. With a set of two sieves, the principal product is typically kept on the screen of the lower sieve.

Separation in milk processing involves the fat and the water in the milk and is aimed at obtaining milk products with a long shelf-life. The separation of fat takes place in two stages: first it is clumped together by '*creaming*' or '*churning*', and then it is physically separated by '*skimming*', pouring, or draining. Draining can occur as a result of gravity alone or as a result of placing pressure on the mixture.

Water can be separated from the milk by '*vaporization*', during which water is removed at the surface (by sun-drying or air-drying) or by the formation of water gas throughout the milk that escapes at the surface (by '*boiling*' down the milk). Vaporization results in a concentration of the nutrients in the milk product.

A second method of separating water from the milk is by draining, in which water, together with dissolved and soluble nutrients, is removed. This kind of separation changes the proportions of certain nutrients in the milk product.

2.2.4 Mixing and shaping

Mixing and shaping are processes that are applied to both cereals and milk. Florets and grain kernels are dry products, which are hard to shape into larger units. Mixing with water or a liquid milk product makes it possible to shape them into larger units. But this is only possible if florets and grain kernels are first granulated or pulverized. Milk, on the other hand, has to be converted into a solid product to make shaping possible prior to drying.

As noted above, grain must be heated to make it digestible. The mixing and shaping of granulated and/or pulverized grain improves the efficacy and efficiency of the heating technology. One option is to produce uniform, composite granules that require the same quantity of heat and that thus make fuel consumption efficient. Examples are bulgur that has been size-graded, '*couscous*', and granulated '*tarhana*'. Another option is to produce a mixture of coarse particles and powdery meal or flour that can then also be mixed into larger units, such as in the production of tarhana. Such small units of tarhana are typically heated by boiling in water. Mixing and shaping can also be done with a specific grade of the fragmented grain. Grain that has been pulverized into meal or flour can be used to make a '*batter*' (more liquid) or dough (less liquid), both of which facilitate more controlled heating. Batter can be heated on a flat or concave surface only, whereas dough can also be heated on a convex surface.

Cereal products made from meal or flour include '*pasta*' (similar product: '*noodles*') and '*bread*'. Shaping into larger units can be done with only the hands, by using a '*mould*', or by using special implements, such as a rolling pin and a table. The shape and size of the individual units is primarily related to efficient processing, efficient storage, and final food preparation. Efficient processing and storage are not fully complementary, and this lack of complementarity can partly explain the variety of shapes that exist for a certain food product. Shaping the individual units into the appropriate quantities before drying makes it possible to obtain the desired dose for food preparation. Shaping grain into larger units also offers the opportunity to express cultural identity (see examples of pasta, bread, and tarhana).

The first use of moulds can be considered to be a way to isolate the food from any nearby soil and fuel. A bread mould can be filled with batter or dough and placed in hot sand. The development of heating installations reduced the use of bread moulds. However, the use of moulds in single-chambered pre-heated '*peel ovens*' still makes sense in this respect (Cappers et al., 2016). Moulds can also be used to standardize the quantity of the food product, which is especially applied in the production of food that is to be traded.

Other products that can be mixed with either cereals or milk include salt, a starter that contains microorganisms, '*rennet*', and fragmented plant parts. Only small quantities of salt are added to cereal-based products to improve the taste. In contrast, substantial amounts can be added to milk products to reduce the water activity and thus the shelf-life of the product. If the product is being fermented, large quantities of salt can only be added once fermentation has finished because salt will lower the water activity and thus inhibit fermentation.

The use of a starter to improve the fermentation process is a tradition in both bread making and souring milk. Although natural fermentation can take place in both dough and milk, the use of a batch of starter with the desired microorganisms is preferred for a controlled fermentation. In order to have access to the desired starter, a small quantity of fermented dough is put aside for the next bread making and then mixed with newly made dough. The '*lactic acid fermentation*' that takes place results in an acid taste, hence the name sourdough. In a similar way, a starter of some concentrated milk with the appropriate microorganisms is used for the inoculation of raw milk. Raw milk is preferably heated before inoculation with the starter, in order to kill the variety of microorganisms that is already present. Kefir globules that are used as a starter have a typical cauliflower-like shape.

Mixing with rennet, an enzyme secreted by the stomach of young mammals, is typically done in milk processing to accelerate the '*curdling*' of casein proteins. Plants can also be mixed, occasionally for improving casein curdling as well, but more often to improve the taste and flavour of the milk product. Such plants seem to be used mainly in the milk-based products '*kashk*' and '*rennet cheese*' and in the cereal–milk-based product tarhana.

2.2.5 Heating and cooling

Heating is applied in the processing of both cereals and milk and serves a multitude of functions. It can be used in cereal processing to remove the awns to reduce the storage volume and to improve sowing, but it can also be applied to facilitate the removal of any remaining chaff. The application of heat is a necessity in the case of cereals to increase their digestibility by humans. Special temperatures are required to optimize biochemical processes in both cereals and milk. These biochemical processes may include germination, fermentation, and enzyme activity. High temperatures kill microorganisms in raw milk and kill harmful animals present in the cereal harvest. They also reduce gluten activity, which facilitates drying of the end product. For example, roasted meal or flour allows for the production of a flat bread that easily dries. High temperatures are also used to kill germinated grain at the end of the malting process.

Variables that play a role in heating are the form of heat transfer, whether or not water or fat is used, and the duration of heating. Dry-heating is done without water; it includes roasting, '*baking*', and '*smoking*'. Wet-heating is done with water; it includes boiling and '*steaming*'. Fat- or oil-based heating is referred to as '*frying*'. Complete heating results in food that is edible, whereas par-heating necessitates further, final heating during food preparation. However, this last stage of heating can be rather short and may only require soaking in warm water (Valamoti, 2011).

Dry-heating is applied to cereals. It involves the processing of unripe grain (frikeh) and ripe grain (galiyya) on a flat, clean surface or, in the case of food processing in hot sand, on a hot stone or by using an oven. Dry-heating is possible both for florets and for grain kernels and for larger food units obtained by mixing granulated or pulverized florets or grain with water or milk. The roasting of florets or whole grains is preferably done by continuously stirring them in hot sand. In this way, heat is transferred to the entire surface of the floret or grain kernel and scorching of the surface is limited. A more advanced version of this method uses sand that has been sieved prior to mixing it with the florets or grain. In this way, the sand will have a consistently smaller diameter than the florets or grain, so that the sand can be quickly and fully separated again after the roasting is finished. If the roasting is done in a concave metal plate, the sand remains pure and can be used repeatedly. In some cases, the florets or grain is soaked in water or milk for a while. Oil can be added to the sand to increase the temperature. High temperatures may result in the '*popping*' (alternative term: '*puffing*') of the florets or grain.

Traditional ovens that are still in use today can be categorized according to the method of isolating the food from the fuel and the ash. One solution is separating the food and the fuel in space, in which a distinction is made between single-chambered ovens (having a single space for fuel, ash, and food) and double-chambered ovens (having a separate chamber for fuel and food or for only food). A second solution is separating the food and fuel in time, in which a distinction is made between pre-heated ovens (baking is done after the oven has been heated) and direct-heated ovens (baking is done while oven is being heated). A further nuance in the typology is obtained by categorizing the oven space in relation to the placement of the food, fuel, and ash. Based on fieldwork in south-west Asia, the Indian subcontinent, and North Africa, five oven types were distinguished: the '*tannur*', the '*tabun*', the '*plate oven*', the '*pan oven*', and the peel oven (figure 32).

The tannur is a single-chambered oven that is heated from the inside and pre-heated when solid fuel is used. Bread is baked in a vertical position. The tabun is a single-chambered oven that is heated with solid fuel from the outside. Bread is baked in a horizontal position. The plate oven is a single-chambered or double-chambered oven with a plate on top that is direct-heated from below. The pan oven is a single-chambered or double-chambered oven with a pan on top that is directly heated from below. The peel oven is a single-chambered, pre-heated oven or a double-chambered oven that is either pre-heated or direct-heated. Bread is baked without a mould or in a mould. The bread or mould is put on a horizontal plate and can be baked twice. Each oven type is characterized by implements that are used for cleaning the oven, putting the food (loaves of bread) into the oven, and taking the food (bread) out of the oven (for a detailed documentation, see Cappers et al., 2016: 745–754).

Wet-heating, which is applied to both cereals and milk, necessitates the use of a container. Before the invention of pottery making, other kinds of containers could have been used, including those made of stone or organic material, such as leather bags and wooden vessels. But even the availability of pottery does not necessarily rule out the use of organic containers. For example, current pastoral Ahaggar Tuaregs (Algeria) prefer to heat their raw milk by putting some hot stones in a wooden bowl rather than by heating it in earthenware pots over a fire, because in the wooden bowl method, the milk remains clean (Nicolaisen & Nicolaisen, 1997).

Boiling can be done with a single container, whereas steaming requires two containers on top of each other. The upper container has a perforated bottom through which steam from the lower container enters the space of the upper container, which is filled with food. Couscous is typically heated by steaming to prevent disintegration of the agglomerated granules.

Cooling to a certain temperature is important if the next step in food processing deals with the activity of microorganisms. In milk, this takes place after heating the raw milk and before inoculating it with a starter. A similar

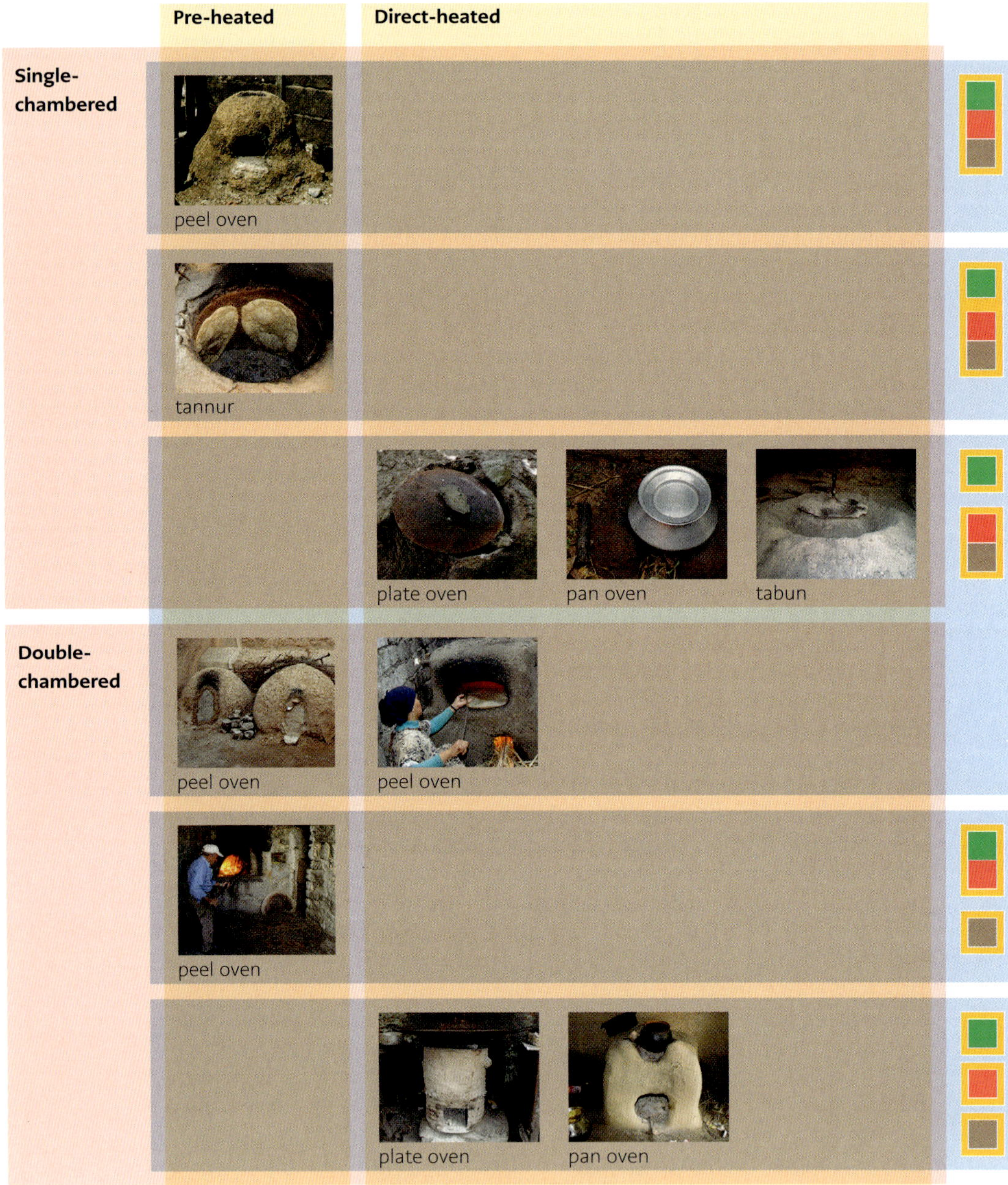

FIGURE 32: Classification of the tannur, the tabun, the plate oven, the pan oven, and the peel oven in relation to the location of the food, the location of the solid fuel and ash, the number of chambers, and the timing of the heating. Pre-heated ovens are only heated prior to food preparation, whereas direct-heated ovens are heated both prior to food preparation (pre-heating) and during food preparation (■: space; ■: food [bread]; ■: fuel; ■: ash).

temperature shift is applied in '*beer*' production, when germinated grain is killed by heating and is then cooled to an optimum temperature for the yeast that is responsible for the '*alcoholic fermentation*'.

2.2.6 Germination

Each seed is a new offspring of a plant, and when the necessary conditions have been met, it can develop into a mature plant. The first stage of growing is designated as '*germination*', or '*sprouting*'. The progenitors of barley and wheat, as well as the first domesticated barley and wheat, are winter annuals that are adapted to the Mediterranean climate. The dispersal of their diaspores takes place before the start of the hot, dry summer. Once buried in the soil, the diaspores remain dormant to survive the period that is not suitable for germination. The breakdown of this stage of dormancy is triggered by favourable environmental conditions that allow the plant to grow into a full-grown plant that is capable of producing new offspring.

Exposure to lower temperatures and an increase in humidity can break down the dormancy and result in germination. Germination can take place while the grain kernel is surrounded by the hulls. The enzymes that break down the cell walls become mobile and convert the starch—which is concentrated in the starchy endosperm—into fermentable glucose.

Today, with an increased interest in alternative and healthy foods, the sprouted grain of barley and wheat is eaten, and it is plausible that such a transformation of a hard, dry fruit into a soft, edible sprout has a long tradition.

2.2.7 Fermentation

Fermentation is a biochemical process that can take place during the processing of both cereals and milk (figure 33). It effects desirable transformations of the food that are related to preservation, texture, taste, and flavour. Two groups of single-celled organisms are used in the fermentation of cereals and milk: bacteria and yeasts, the latter group being true fungi. Both types of organisms obtain their cellular energy through the fermentation of carbohydrate.

Carbohydrates are molecules that consist of carbon (C), hydrogen (H), and oxygen (O). An important group of carbohydrates are sugars (saccharides), whose basic unit consists of three carbon atoms. Sugars containing this number of carbon atoms are called monosaccharides. Sugars consisting of six carbon atoms are called disaccharides.

Glucose is a monosaccharide. It is produced by photosynthesis in the green tissues of plants. By this process, the energy of sunlight is converted into chemical energy that is subsequently used to combine a molecule of water with a molecule of carbon dioxide. This chemical energy can be released again by breaking the glucose into, for example, water and carbon dioxide. Glucose is the basic organic molecule that is used to synthesize all other kinds of molecules that are necessary, including other carbohydrates, proteins, lipids, and nucleotides. The specific minerals that are needed for synthesizing all these molecules are obtained by the uptake of soil water and mineral content.

Glucose is soluble in water and can be transported in plant vessels to, for example, the developing grain kernel, where it is stored as a source of energy. Because the concentration of glucose determines the water activity, it is converted into starch by linking many glucose molecules to each other. Each linkage of two monosaccharides is a dehydration reaction. Starch, being a polysaccharide, has no impact on the water activity.

The carbohydrate in milk differs from that in a grain kernel. Whereas the carbohydrate in grain, termed starch, is a polysaccharide entirely made up of

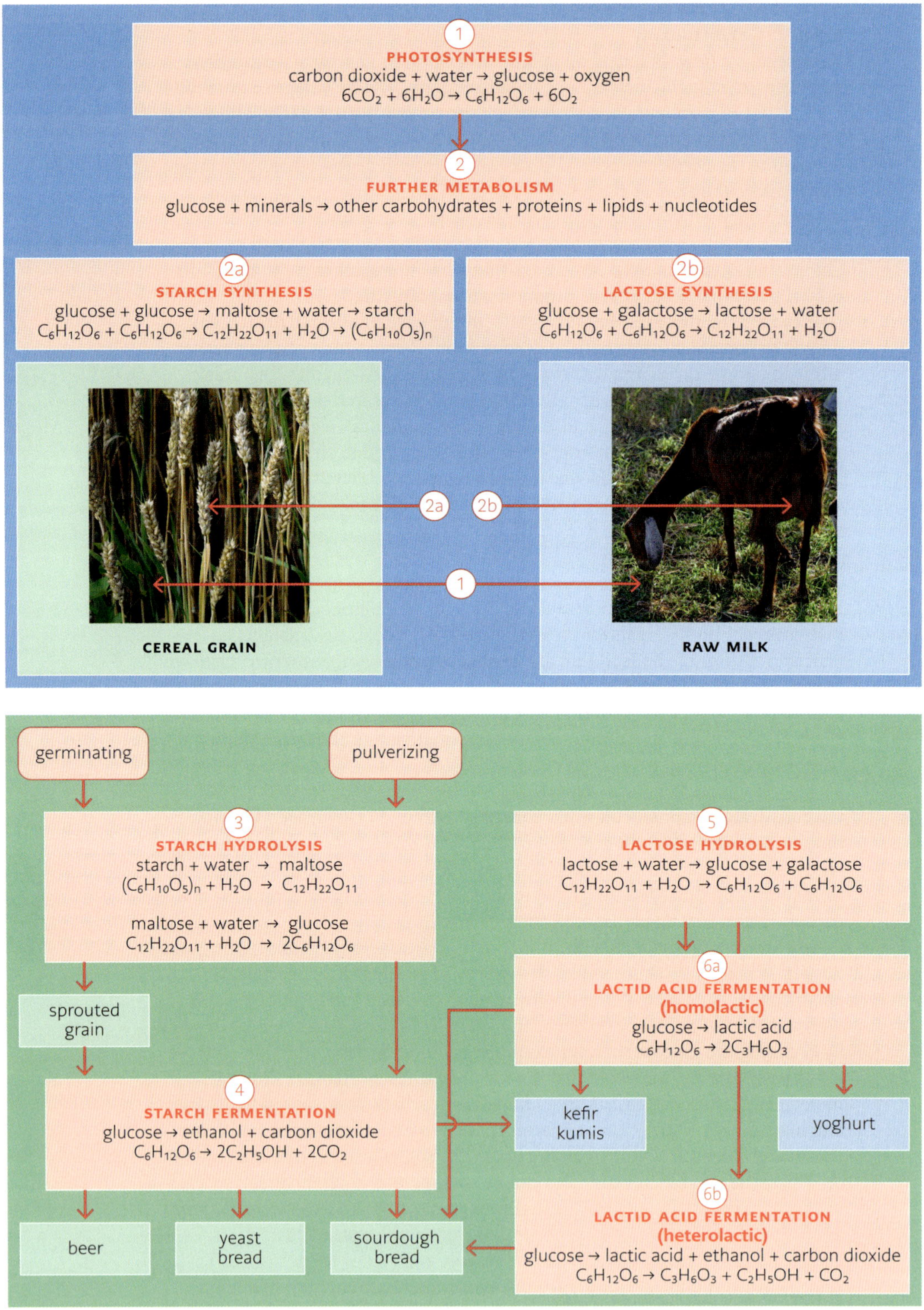

FIGURE 33: Metabolism of carbohydrates as it relates to the production of selected grain products and milk products. Only the main, overall equations are shown.

glucose, the carbohydrate in milk, termed milk sugar (or lactose), is a disaccharide that consists of two different monosaccharides: glucose and galactose.

Glucose can be fermented by both bacteria and yeast, whereas lactose can only be fermented by bacteria. This implies that the starch present in the grain endosperm first has to be converted into fermentable glucose by the uptake of water, a process termed hydrolysis. This starch hydrolysis takes place during germination, but it can also take place when meal or flour, which predominantly consists of starchy endosperm, is mixed with water. Lactose hydrolysis is a similar process; it is necessary to convert lactose into glucose and galactose to thereby make it possible for bacteria and yeast to perform further fermentation.

Two different kinds of fermentation occur in cereals and milk: lactic acid fermentation (LA-fermentation) and alcoholic fermentation. Both kinds of fermentation occur naturally, because the single-celled bacteria and yeast easily enter fragmented grain and milk. Traditionally, the same yeast was used for beer and bread making. Today, different strains of bacteria and yeast are used in the production of beer and bread in order to optimize the respective biochemical processes. A variety of lactic acid bacteria normally occur in milk that is exposed to air, and these bacteria cause spontaneous fermentation (Stanier et al., 1981). The storage, transport, and processing of milk in containers that are reused without having been thoroughly cleaned may contribute to the formation of colonies of bacteria that are needed for the desired fermentation. In order to control the quality of the fermentation, raw milk is heated to kill these unwanted microorganisms and is subsequently mixed with some yoghurt made the previous day or with dried yoghurt that may have been kept for some time. In the Little Pamir region of north-eastern Afghanistan, the small quantity of yoghurt culture that is used for mixing with new milk is called '*khumeth*' (Oudenhoven & Haider, 2015).

In LA-fermentation, glucose or lactose is converted into lactic acid (milk acid). Two different biochemical pathways are possible: homolactic fermentation and heterolactic fermentation. In homolactic fermentation, lactic acid is the prevailing product that is produced. In heterolactic fermentation, a mixture of lactic acid, ethanol (alcohol), and carbon dioxide is produced. In ethanol fermentation (also termed alcohol fermentation), glucose is converted into ethanol and carbon dioxide.

Certain food products involve only LA-fermentation, other products involve only alcoholic fermentation, and still other products involve both. The conversion of raw milk into sour milk, such as yoghurt, is the result of only LA-fermentation. During the production of beer and yeast bread, only alcoholic fermentation takes place, but apart from that similarity, the processes for beer and yeast bread differ. In the production of beer, the aim is to convert starch into alcohol as much as possible; the carbon dioxide is a by-product that is released. In contrast, in the production of yeast bread, the aim is to produce a sufficient quantity of carbon dioxide, because this gas becomes trapped in the gluten network and thus produces a raised (leavened) dough. The ethanol is a by-product, and it escapes from the dough by vaporization. In the processing of sourdough bread and certain milk products, such as kumis (originating from Turkey) and kefir (originating from the Caucasus), both LA-fermentation and ethanol fermentation take place. This results in a rather sour taste in the case of the bread and a low alcohol content in the case of the kumis and kefir. Because the gluten are weakened by protein-digesting enzymes and because carbon dioxide production by yeast is inhibited due to the predominance of the bacteria, sourdough bread does not rise very well (McGee, 2004).

2.3 Cereal processing

2.3.1 Classification

Cereal products can be characterized based on their ingredients, the kinds of processes that are applied, and the sequence of the processes. In some foods, the duration of a process also contributes to the food's identity (figures 34 and 35). Heating, for example, can be partial or total. The distinction between crop processing and food preparation is not always clear-cut. This is particularly the case when the harvest is processed on a large scale and includes the removal of the chaff through par-heating, because such processing improves the digestibility as well.

Cereal product	**Grain**			**Other ingredients**				**Short description**
	Whole	***Granules***	***Meal/flour***	***Water***	***Salt***	***Yeast***	***Spices***	
galiyya	○							roasted grain
frikeh	○	(○)						roasted unripe grain (sometimes granulated)
melthuth/tchicha		○						roasted granulated grain
sprouted grain	○			○				germinated grain
koliva	○			□				boiled grain mixed with a variety of ingredients
couscous		○	○	○	(○)			granulated–pulverized grain mixed with water (and salt), made into granules of uniform size
bulgur		○		□				par-boiled granulated grain
polenta		○		□				boiled granulated grain
mirmiz (barley) frik (wheat)		○	○	□	○			par-boiled and roasted granulated–pulverized unripe grain
beer		○		○		○	○	alcoholic beverage from coarsely ground (preferably hulled) grain, water and yeast
pasta		○	○	○	○			shaped mixture of finely granulated or pulverized grain, water and salt; eventually par-roasted or par-boiled
bread			○	○	○	○		shaped dry-heated mixture of pulverized grain, water, salt and yeast

FIGURE 34: Ingredients of some traditional food items made from barley and wheat (□ = used for heating; (○) = optional use).

Cereal product	Hull tightness	Fragmentation: Threshing	De-graining	Granulation	Wetting/drying/heating: Pulverization	Cold-drying	Dry-heating	Wet-heating	Mixing	Germination	Fermentation	Short description
galiyya	H	1	2				3					roasted grain
	N	1					2					
frikeh	HN	2		(3)			1					roasted unripe grain (sometimes granulated)
melthuth/tchicha	H	1	2	4			3					roasted granulated grain
	N	1		3			2					
sprouted grain	N	1								2		germinated grain
koliva	H	1	2					3	4			boiled grain mixed with a variety of ingredients
	N	1						2	3			
couscous	H	1	2	3					4			granulated–pulverized grain mixed with water (and salt), made into granules of uniform size
	N	1		2					3			
bulgur	H	1	4	5		3		2				par-boiled granulated grain
	N	1		4		3		2				
polenta	N	1		2				3				boiled granulated grain
mirmiz (barley) frik (wheat)	HN	3		5	5	2	4	1				par-boiled and roasted granulated–pulverized unripe grain
beer	H	1		5		4	3	7	6	2	8	alcoholic beverage from coarsely ground (preferably hulled) grain, water and yeast
pasta	H	1	2	3		5	5	5	4			shaped mixture of finely granulated or pulverized grain, water and salt; eventually par-roasted or par-boiled
	N	1		2		(4)	(4)	(4)	3			
bread	H	1	2		3		6		4		5	shaped dry-heated mixture of pulverized grain, water, salt and yeast
	N	1			2		5		3		4	

FIGURE 35: Sequencing of food preparation of some traditional food items made from barley and wheat. Mixing is said to occur when pulverized grain is included. Processes that are considered as optional are placed in parentheses. Polishing and pearling as a possible method of fragmentation are not included. Separation, which occurs after threshing and de-graining, is not indicated (H: hulled; N: naked).

2.3.2 Harvesting and threshing

In order to explore the traditional food items that it is possible to make from barley and wheat, it is necessary to take into account the morphological differences between barley and wheat. Different harvesting methods can be applied; the combination of a particular harvesting method, rachis brittleness (whether wild or domesticated), and ripening stage (whether unripe or ripe) determines the kind of harvesting unit. Plausible methods for harvesting wild and domesticated barley and wheat, together with the corresponding harvesting units, are presented in figure 36.

Picking up and beating apply only to wild cereals. Picking up involves picking up the dispersal units (diaspores) from the ground surface. Beating involves the touching of the spikes, for example with a swinging basket or a

Harvesting method	Harvesting unit			
	Wild		*Domesticated*	
	Unripe	*Ripe*	*Unripe*	*Ripe*
picking up		rachillate triplet (B) rachillate spikelet (W)		
beating		rachillate triplet (B) rachillate spikelet (W)		
plucking		rachillate triplet (B) rachillate spikelet (W)		culm + spike
stripping off	culm + spike		culm + spike	culm + spike
cutting	culm + spike		culm + spike	culm + spike
uprooting	root + culm + spike		root + culm + spike	root + culm + spike

FIGURE 36: Relation between harvesting method and harvesting unit of barley (B) and wheat (W), taking into account the rachis brittleness (wild versus domesticated) and maturity of the grain (unripe versus ripe). Classification of harvesting methods based on Anderson and Sigaut (2014).

combination of a basket and a paddle or stick, by which ripe dispersal units are collected in the basket.

Plucking applies to either wild or domesticated cereals. Plucking involves individually breaking off the ripe dispersal units (in wild barley and wheat) or ripe spikes (in domesticated barley and wheat) from the plant by hand.

Stripping, cutting, and uprooting apply to unripe wild cereals or unripe or ripe domesticated cereals. Stripping off spikes from the culm can be done with the bare hands or, more conveniently, using the pressure of a tool, such as the *mesorias* that is still used for the harvesting of hulled wheat in northern Spain (Peña-Chocarro, 2014). Cutting is a harvesting method that makes use of a sharp tool with which the culms are cut. Contemporary cutting implements used for the harvest of cereals can be divided into four main types: harvesting knife, sickle, scythette, and scythe. The harvesting knife consists of a short, straight blade that is fitted to a short handle and is adapted for the cutting of individual culms. The sickle is suitable for cutting several culms together. Collecting of the culms can be done with a curved blade with a more or less crescent-shaped cutting edge that is fitted to a short handle. This type is the most common. In Cambodia, a different kind of sickle is used, whereby the culms are collected by means of a crescent-shaped branch and then cut with a cutting blade that is comparatively small. The scythette consists of a long, curved blade that is fitted to a medium-sized handle with a single grip. One hand is used for holding the scythette, and the other is used for holding the (mowing) hook. The scythe consists of a long, curved blade that is fitted to a long handle having either one or two grips. Both hands are used for holding the scythe during mowing (for a detailed documentation, see Cappers et al., 2016: 228–240). Finally, uprooting is done by pulling at the whole plant, during which part of the roots remain attached to the culm.

Taking into account these possible harvesting methods, three different harvesting units can be distinguished in both barley and wheat: a rachillate triplet, a rachillate spikelet, and a spike with at least part of the culm. Rachillate triplets and rachillate spikelets are obtained by harvesting ripe wild barley and wheat by picking them up from the soil, by beating against the spikes, or by plucking ripe diaspores by hand. Whole spikes with part of the culm are obtained if harvesting is done by stripping, cutting, or uprooting an unripe wild cereal or unripe or ripe domesticated cereal. Entire culms with part of the roots are obtained if the cereal is uprooted.

Threshing is the first stage of fragmentation that is applied when the harvesting unit contains a spike. This fragmentation process is aimed at separating the harvest into a principal product (termed '*threshing unit*') and a by-product (termed '*threshing remains*'). The fragmented but unseparated mixture can be designated with '*threshing product*'. Similar concepts can be used in relation to other methods of fragmentation, such as polishing and pearling (see Glossary). Irrespective of the maturity of the grain, the threshing unit of hulled barley is a floret, that of hulled wheat is a rachillate spikelet, and that of naked barley and wheat is a grain kernel (see also figure 37).

2.3.3 Improving edibility and shelf-life

Further fragmentation is aimed at improving the edibility and the shelf-life. The edibility is improved by the removal of inedible parts of the harvesting unit and the threshing unit. The shelf-life is improved by the removal of the outer part of the grain kernel, which contains most of the lipids.

In order to judge the need for this further fragmentation and to describe the processes with terms that are correct and explanatory, it is important to take into account the morphology of the harvesting units of the wild cereals and the threshing units of the domesticated cereals. A clear understanding of this stage of crop processing and its implications for further food preparation will benefit from the use of terms that are both correct and meaningful. The harvesting unit of wild barley is a rachillate triplet; that of wild wheat is a rachillate spikelet; and that of domesticated hulled wheat is also a rachillate spikelet. And the rachillate triplet and the rachillate spikelet clearly differ in their morphology (see section 1.2).

Although '*de-hulling*' is an established term to designate the fragmentation of rachillate spikelets (Cappers et al., 2016), it is not an adequate term to describe the process concerned. The term de-hulling is applied to rachillate spikelets of wild and domesticated wheat to refer to a process in which both the rachis segments and the hulls—glumes, lemma, and palea (including those of the sterile florets)—and the awns are removed. But the term de-hulling ignores the removal of the rachis segment.

The removal of inedible parts from a rachillate triplet that is harvested from wild barley stands is a special case. Here, threshing is not applied, and a special treatment is necessary that has no parallel in the processing of either unripe wild barley, domesticated barley, or wheat.

The discussion above underlines the need for specific words that designate the processing of rachillate triplets and rachillate spikelets. The most obvious solution seems to be to use a term that is related to the principal product ('*floret*' in barley, '*grain kernel*' in wheat) instead of a part of the by-product (*hulls*). The alternative terms suggested here are therefore '*de-floretting*', for the fragmentation of the rachillate triplets, and '*de-graining*', for the fragmentation of the rachillate spikelet. De-floretting is thus the process that is aimed at isolating the floret from the inedible parts, whereas de-graining is the process that is aimed at isolating the grain kernel from the inedible parts (figures 37 and 38).

The proportion of non-edible parts in a rachillate triplet and a rachillate spikelet is high. Therefore it is necessary to remove most or all of these inedible parts in order to obtain a unit that is suitable for food preparation. The rachillate triplet, which is harvested from wild barley, is fragmented and separated into a fertile floret and a by-product that consists of a rachis segment; six spiny, awned glumes; two sterile florets; and the awn of the fertile floret.

Hulled				Naked
Ripe wild barley	***Partly unripe wild barley/domesticated barley***	***Ripe wild wheat***	***Partly unripe wild wheat/domesticated wheat***	***Domesticated barley/domesticated wheat***
harvesting	harvesting	harvesting	harvesting	harvesting
rachillate triplet	culm and spike	rachillate spikelet	culm and spike	culm and spike
	threshing		threshing	threshing
	culm rachis glume ③ (sterile floret) awn (lemma)		culm ④ awn (lemma)	culm rachis glume ⑥ lemma & palea
	floret		rachillate spikelet	grain kernel
de-floretting		de-graining (de-hulling)	de-graining (de-hulling)	
rachis segment glume ① sterile floret awn (lemma)		rachis segment glume ② sterile floret lemma & palea	rachis segment glume ⑤ sterile floret lemma & palea	
floret		grain kernel	grain kernel	
pearling	pearling	polishing	polishing	polishing
lemma & palea bran germ	lemma & palea bran germ	bran germ	bran germ	bran germ
pearled grain	pearled grain	polished grain	polished grain	polished grain

FIGURE 37: Processing of wild and domesticated barley and wheat. In hulled cereals, the harvesting unit differs between partly unripe and fully ripe infructescences. Processes are shown in blue and the plant units are shown in green, subdivided into dark green for plant units contain a rather large amount of inedible plant parts and light green for plant units that are edible after heating. By-products of the fragmentation processes are shown in red. The numbers of the by-products of threshing, de-floretting, and de-graining correspond with those in figures 14 and 38. Separation processes are not indicated. The culm part varies in length and may include roots as well.

The rachillate spikelet that is obtained during the harvesting of wild wheat or the threshing of domesticated hulled wheat is fragmented and separated into grain kernels and a by-product that consists of a rachis segment, two broad glumes, and six lemmas and paleas. The lemma and palea belong to both fertile and sterile florets.

Awns of rachillate triplets (from both the glumes and the lemma of the central floret) and rachillate spikelets (from one or more of the lemmas) are rather long and stiff. Their presence prohibits efficient transport, storage, and sowing. Therefore, de-awning can be applied as an intermediate process to reduce the size of the awns and thereby improve the compaction of the storage units.

Unfortunately, there is a lack of clear documentation of traditional methods

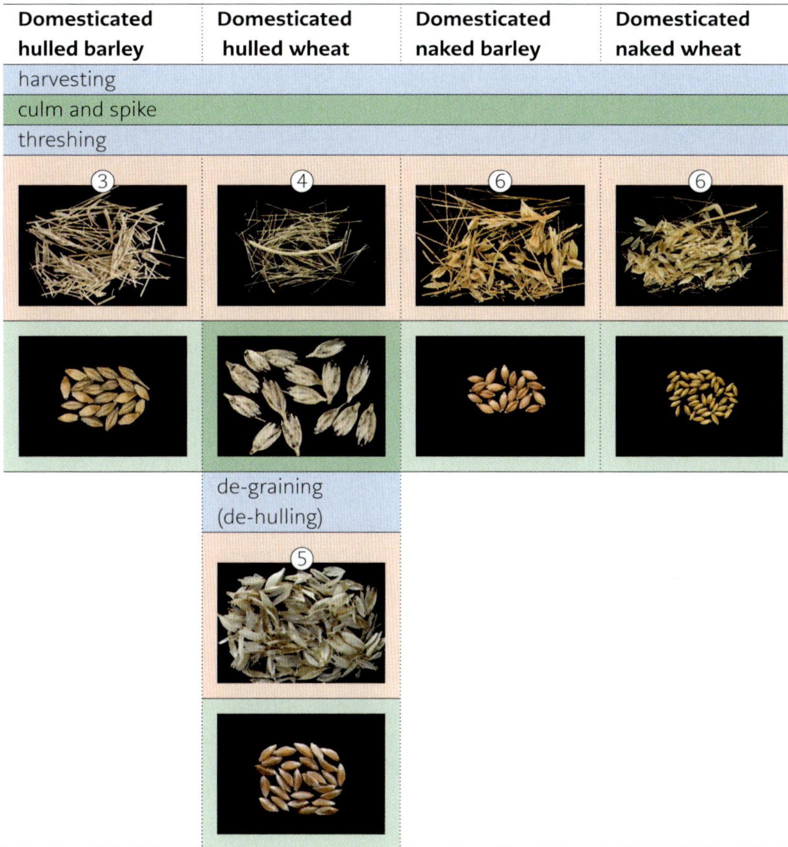

FIGURE 38: Principal products and by-products of ripe hulled and naked domesticated barley and wheat. From left to right: hulled 2-row Barley (*Hordeum vulgare* ssp. *distichon*); hulled wheat: Emmer (*Triticum turgidum* ssp. *dicoccon*); naked 2-row Barley (*Hordeum vulgare* ssp. *distichon*); naked Hard wheat (*Triticum turgidum* ssp. *durum*). Each example shows the quantity that is produced through the manual fragmentation of a single spike. The numbers of the by-products corresponds with those in figure 14 and 37. The photos are to scale.

of de-graining (or, as it is termed elsewhere, de-hulling) rachillate spikelets of wheat in historical and ethnographic sources. Today, hulled wheat is grown on a limited scale, and it is only de-grained if grown for human consumption. Nevertheless, some methods are still in use, and these illustrate different pathways. De-graining without the use of heat is still practised in the processing of Einkorn (*Triticum monococcum* ssp. *monococcum*) in the western Rif region in northern Morocco. Small quantities of rachillate spikelets are rubbed with a circular piece of cork or beaten with a mallet (which may also be used for washing clothes). Large quantities are de-grained by pounding the grain in mortars or by grinding, in which sheets of cork are used to prevent fragmentation of the grain (Peña-Chocarro et al., 2009). Experimental de-graining of Einkorn (*Triticum monococcum* ssp. *monococcum*), Emmer (*Triticum turgidum* ssp. *dicoccon*), and Spelt (*Triticum aestivum* ssp. *spelta*) shows that pounding is much more efficient than grinding (Meurers-Balke & Lüning, 1990). Rachillate spikelets of emmer that were de-graining with the aid of a saddle quern produced a mixture of unprocessed rachillate spikelets, florets, grain kernels, granulated grain, and pulverized grain. Such a mixture is not easily separated into fractions that can be further processed. De-graining rachillate spikelets by pounding produced a mixture of unprocessed rachillate spikelets and grain kernels (in the case of einkorn) or a mixture that also included some florets (in the case of emmer and spelt). Separating the unprocessed spikelets and florets from the grain kernels and repeating the pounding will produce the desired quantity of grain kernels.

The use of fire is often mentioned in historical sources, but understanding the specific intention is not always easy from the descriptions. Experiments indicate that dry-heating prior to wet-pounding is not necessary (Meurers-Balke & Lüning, 1990; Samuel, 2000; for a critical review, see Peňa-Chocarro & Zapata, 2014). Nevertheless, dry-heating and wet-heating are still applied for de-graining rachillate spikelets. In both cases, the heating is partial, and this stage of crop processing of grain interferes with food preparation. Because of the partial heating, a final, short heating during food preparation is necessary.

Partial dry-heating through roasting ('*parching*') is still used in the preparation of frikeh (see section 3.2.3). Although nowadays frikeh is made from naked wheat, the Arabic word *afrakah*, which means 'rubbing between hands', refers to the fact that frikeh originally involved processing hulled cereals (Musselman & Al-Mouslem, 2001). Partial wet-heating by par-boiling prior to de-graining between grinding stones is still practiced in the processing of Einkorn (*Triticum monococcum* ssp. *monococcum*) in the Black Sea area of Turkey (Cappers et al., 2016). The par-boiling results in an increase in the grain size, due to the absorption of water, and in gelatinization of starch. The expanding grain disconnects from the lemma and the palea, with the result that both pieces of hull can be more easily removed. And the gelatinized starch results in a firm grain kernel that is more easily granulated or pulverized. This kind of de-graining is also applied to rice (*Oryza sativa*).

If florets and grain kernels are to be used for making dry food with a long shelf-life, it is desirable to also remove the bran and the germ. Both parts contain a high concentration of lipids in particular, and removing these parts prevents the drying food becoming mouldy. Sun-drying, in particular, is a critical stage in this respect.

The fragmentation process that deals with the removal of bran and germ is called '*polishing*'. Traditionally, this is done by pounding wet grain kernels. While pounding a quantity of wet grain kernels, the grain kernels rub against each other and small flakes of bran and germ become separated. The bran fragments and germ can be separated from the polished grain after drying. The difference in shape and size does not matter much with this kind of technique. Using grinding stones instead of pounding, in contrast, places higher demands on the uniformity of the grain kernels. The surface of the grinding stones should be of a high quality, and their distance should be adjusted with much precision to prevent granulation or pulverization. Polishing rather uniform grain kernels, such as those of Einkorn (*Triticum monococcum* ssp. *monococcum*), Emmer (*Triticum turgidum* ssp. *dicoccon*), or Spelt (*Triticum aestivum* ssp. *spelta*), will be more efficient than polishing those of Hard wheat (*Triticum turgidum* ssp. *durum*) or Bread wheat (*Triticum aestivum* ssp. *aestivum*), because in these latter two taxa, grain kernels from a single spikelet differ markedly in size. To reduce loss of the starchy endosperm, modern polishing machines separate grain kernels into different size classes. Then each size class is polished with rotating rollers with the appropriate space between them.

Separate de-graining and subsequent polishing of a floret is difficult to achieve because hulls (lemma and palea) are tightly connected with the grain kernel. The tight connection between hull and grain kernel is favourable for storage, because the grain kernel is protected by the hull, whereas the hull barely increases the volume of the storage unit. If the hull is removed, this usually also results in the removal of the bran and the germ. This simultaneous removing of hull, bran, and germ is designated as *'pearling'*, and the grain kernel that is obtained after removing the particles of hull, bran, and germ is designated as *'pearled barley'*. There seems to be no specific name for the by-product of pearling.

2.3.4 Granulating and pulverizing

The final fragmentation of the floret and the grain kernel is aimed at maximizing the available fuel and at facilitating the mixing and shaping of the fragmented florets or grain into other food items. If this kind of fragmentation is done by pounding, the fragments usually differ in size, ranging from coarse granules to powdery particles. The mixture is then separated into different size classes to make efficient use of fuel possible. The granulated or pulverized florets or grain kernels can also be mixed with water for the preparation of such food items as couscous, pasta, or bread, or with milk for the preparation of tarhana.

There is no need to remove the hull from the floret in order to obtain a food item that is edible. Because of the tight connection between the hull and the grain kernel, both are jointly fragmented to small units. If a meal sieve is available, the coarse fragments of the hull can be separated from the meal. The tradition of pulverizing florets and further processing them into a food item may even lead to an appreciation of the specific texture that is obtained. For example, people from the Greek island of Melos (in the Aegean Sea) consider the presence of coarse particles of hulls in their barley *'rusk bread'* as a guarantee of its taste (Procopiou, 2014).

2.4 Milk processing

2.4.1 Classification

Although the principal compounds and basic processes are limited, it is possible to produce a large variety of milk products from raw milk, even without taking into account the impact of plant material consumed by the lactating animal and the breed of animal that is milked. The milk products available today are usually recognized as belonging to one of four categories: yoghurt, cream, butter, or cheese. However, the identification and classification of a particular milk product to one of these categories is not always straightforward, because criteria used for their identification are not unequivocal. Identification and classification are further obscured by linguistic problems.

A similar milk product can be designated with different names after its geographical spread into areas with different language families, and vice versa. For example, the Turkish word *'yoğhurt'* is used for fermented milk, whereas the Arabic word from the same root, *'yëgo(r)t'*, is used for *'curds'* obtained by draining *'buttermilk'* (Ferdinand, 1993). But also within a particular language, a certain term can be used for different milk products. The Turkish term *'ayran'* is used for buttermilk, but also for yoghurt that has been diluted with some water to make it easier to drink. Furthermore, the meaning of a term can change over time (see, for example, the different meanings of the word '*kashk*' in Aubaile-Sallenave [1994] and in section 3.3.1).

With respect to the conversion of perishable milk into milk products with a long shelf-life, it is of interest to look at what kinds of '*dried milk products*' and dried cereal–milk products can be made (figures 39 and 43). A description of dried milk products that are still produced today can be used as a frame of reference for the reconstruction and interpretation of former milk processing, which became a challenge once a surplus of milk started to be produced. In order to obtain a better idea about the milk products that could, potentially, have been produced in the remote past, it is of interest to see what can be made by means of basic processes. In this respect, it is of interest to note that the heating of milk is not essential in milk processing, although it reduces the risk of spoilage and may improve the quality of the milk product.

A systematic investigation into the different possible pathways of milk processing has led us to suggest all the possible permutations of dried milk products, as detailed in figure 39. Some of these products could not be traced back in descriptions of traditional milk processing but result in a similar kind of milk product (that is, a '*non-rennet cheese*'); these have been labelled with the word '*kashk*' pending further specification.

The conversion of raw milk into milk products with a long shelf-life is important not only for families who milk their animals at home, but also for shepherds who take care of animals in pastures during the summer period. When shepherds move with their flock to another area or return home, the dried milk products they have produced should remain unspoiled and be transportable. In this context, for the Little Pamir region in north-eastern Afghanistan, Edelberg and Jones (1979) mention butter and cheese and Oudenhoven and Haider (2015) mention butter and dried buttermilk (local term: *tukhp*).

When the main focus is on milk products with a long shelf-life, any processes applied to the milk have to be evaluated for their contribution to the preservation of these products. The shelf-life of a milk product is determined by its acidity and its water activity. A lowering of the acidity is achieved through the fermentation of the milk sugar. And the lowering of the water activity is realized by solidifying and drying the milk and by adding salt. Converting the liquid milk into a '*solid milk product*' is possible by clumping of the milk fat and by coagulation of the milk proteins. Clumped fat and coagulated proteins can be subsequently separated from the liquid part. Drying is the final process in improving the shelf-life of a milk product. During drying, most of the moisture is removed, and once this has been accomplished, the dried milk product can be stored for a considerable length of time.

In identifying and classifying a particular milk product, it makes sense to take into account the composition of the milk product and the different processes that are applied. The main processes are related to four different ingredients of the milk: milk sugar, fat, proteins and water. This implies that the processes involved in the production of a specific milk product will also determine its final composition (for the composition, see also section 1.3.2).

2.4.2 Fermentation of milk sugar

Fermentation in milk is done by lactic acid bacteria that are naturally present in the raw milk, which implies that some fermentation will usually take place

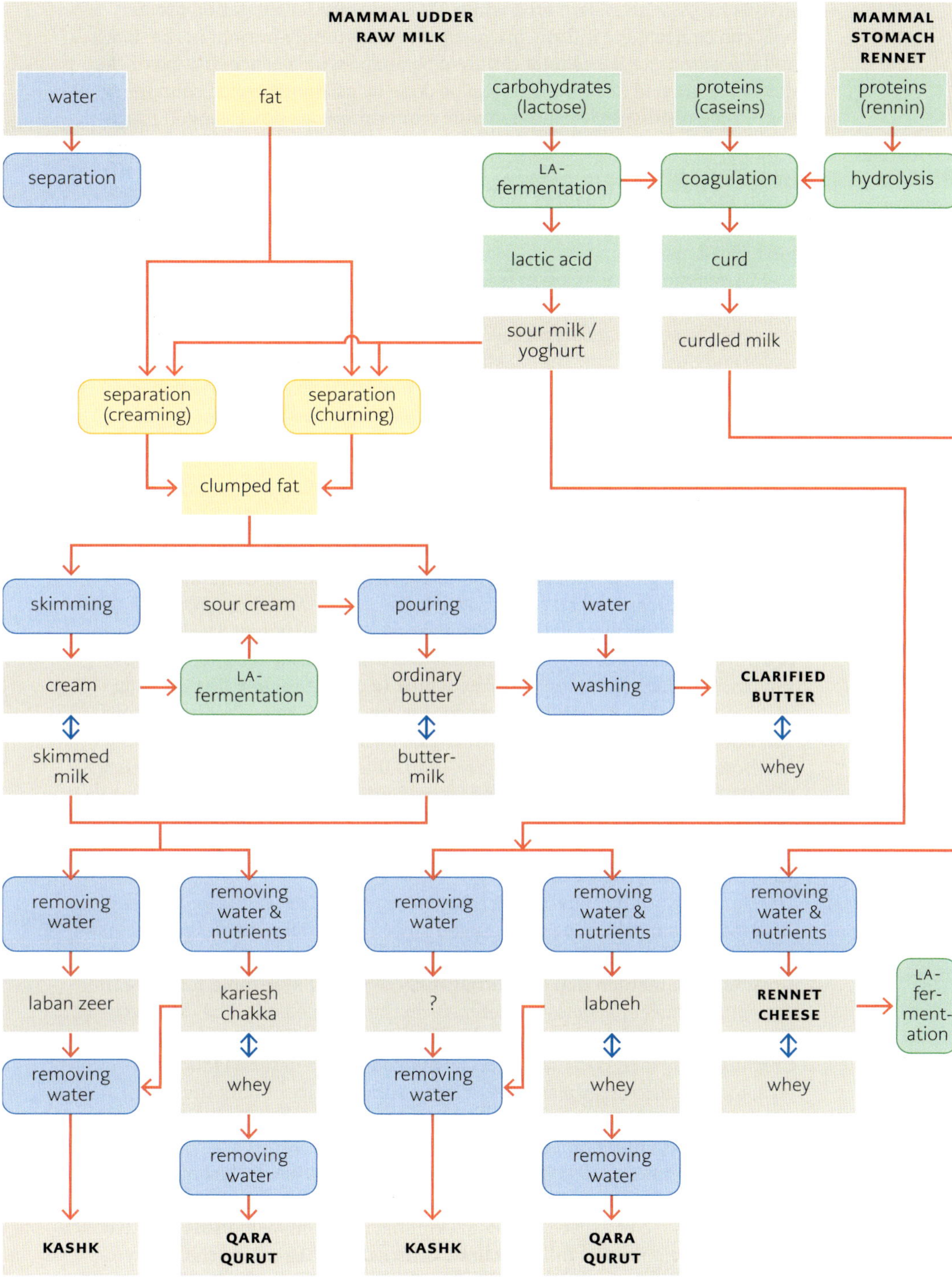

FIGURE 39: Simplified model showing the basic processes involved in the production of milk products having a long shelf-life. Heating as a possible process is not indicated. Storable products are presented in boldfaced capitals. Single arrows indicate the conversion into a new product; double arrows indicate the separation into a principal product and a by-product (LA-fermentation = lactic acid fermentation).

in milk that has been kept for a while. To control the fermentation process, milk can be heated to kill any microorganisms naturally present in the milk, and the now pasteurized milk can then be mixed with a concentration of the desired lactic acid bacteria. This can be done by using unwashed containers containing residues of fermented milk from a previous processing. It is also possible to supplement (inoculate) the raw milk with a batch of fermented milk from a previous processing, which can be either yoghurt or buttermilk that is made from fermented milk. Adding a batch of fermented milk is thus similar to the use of a small piece of fermented dough (sourdough) for the next bread making.

Microorganisms that are present in raw milk can also be killed by adding some salt. Salt is usually added once fermentation has finished. If it is added at the very start of the milk processing, as is, for example, traditionally done in Egypt and Sudan during the making of the soft cheese *'gibneh beydah'* (Dagher, 1991), all of the bacteria that are naturally present are killed and no lactic acid fermentation occurs.

Yoghurt is sour milk with a semi-solid structure. It is characterized by low acidity, low sugar content, and some coagulation of proteins. Despite its acidity and its lower moisture content, yoghurt's shelf-life without being kept cool is limited to several days, and a surplus of yoghurt therefore has to be further processed into other milk products to keep it edible. Some other words for sour milk or yoghurt are *'pai'* (Afghanistan and Tajikistan: Oudenhoven & Haider, 2015); *'mast'* (western Iran: Mortensen, 1993; Watson, 1979); *'rôb'* (Qatar: Ferdinand, 1993; Sudan: Wilson, 2017); *'zabady'* (Egypt: Dagher, 1991); and *'rayeb'* (Maghreb: Dagher, 1991). The word *'laban'* may refer to raw milk (Sudan; Wilson, 2017); sour milk or yoghurt (e.g. Gulf region: Dagher, 1991; Weir, 1990; Arab world: Wilson, 2017); or buttermilk (Qatar: Ferdinand, 1993). The word *'laban'* is similar to the word *'lben'*, which is used for buttermilk in north-west Africa (Maghreb: Dagher, 1991). This linguistic similarity suggests that the use of fermented milk for making butter and buttermilk is a widespread method of processing raw milk.

2.4.3 Solidifying of fat

The clumping of fat can be achieved by two methods: creaming and churning. Both methods result in the clumping of small fat particles into larger ones, and both can be considered as the first stage of separating fat from raw milk. Creaming is a method that occurs naturally if raw milk is stored for some time in a container. Fat particles concentrate on the surface under the influence of buoyancy and form a cream layer that floats on the now *'de-fatted milk'*. Creaming may take 12–24 hours, during which period natural lactic acid fermentation also takes place, resulting in some acidity of both the cream and the de-fatted milk. The second stage of separation is the removal of the floating cream; this is called *'skimming'*.

A quicker method of fat solidification is churning, in which the raw milk is sloshed. Milk sloshing can be done in all kinds of devices. Traditionally, churning is done in a dried mammal skin that has been fashioned into a bag. If only a small quantity of milk has to be churned, the bag can be shaken by holding it between one's hands and agitating it (see, for example, Nicolaisen & Nicolaisen 1997: figure 10.11). The same authors also mention the attachment of a skin bag filled with some milk to a camel saddle, after which the camel is ridden for some time at high speed. A similar method is also mentioned for the nomads in Tibet, who fastened skins or stomachs filled with milk to pack animals on the move (Hermanns, 1949). Larger skins can be suspended and agitated by one or two hands (see atlas section: figures 84 and 85) or rolled back and forth over the ground (see, for example, Frederiksen, 1996: figure 3.17) or over the knees (see, for example, Edelberg & Jones 1979: photo 63). According to Edelberg and Jones, a churning skin will last for a month, and about six specimens are needed during the summer period. Large skins can be positioned with the opening on top, which allows for pounding with a wooden paddle.

A funnel can be used for pouring the milk or cream into the skin. The skin is then inflated by blowing air into the skin, after which the opening is closed. During churning, extra water is added if necessary. Once the fat has clumped into large units, the butter clumps are taken out by hand, and then the buttermilk is poured out. The skin is cleaned by adding water, which can subsequently be added to the buttermilk.

The churning of milk is also done in large fruits of the Bottle gourd (*Lagenaria siceraria*) hanging down from a tripod, for example, by the Bani Mughaid Bedouins of Asir, in south-western Saudi Arabia (Mauger & Mauger, 1986: photo C12), and the Ayr Bedouins in Niger (Nicolaisen & Nicolaisen, 1997). Nowadays, a metal container is also used for churning (see atlas section: figure 86). In Qatar, an aluminium churning device was documented and sampled as far back as the 1950s (Ferdinand, 1993).

Butter can be made directly from unfermented milk, but this is labour intensive. The use of fermented milk and cream shortens the process (Oudenhoven & Haider, 2015). Because the fat content of raw milk is rather low, the amount of butter obtained from the creaming of raw milk is limited. Collecting cream from raw milk and separating the fat from buttermilk by churning results in larger quantities for the volume that has to be handled during processing.

The separation of cream and butter from, respectively, '*skimmed milk*' and buttermilk can be regarded as a first step in removing water. The now de-fatted milk is, however, still rich in suspended nutrients (fat and casein protein) and dissolved nutrients (milk sugar, '*whey*' proteins, and minerals). Cream, butter, de-fatted milk and cheese still need to be further dehydrated to enable long-term storage.

The removal of cream is done by skimming; the English term '*skimmed milk*' refers to this process, whereas the Dutch term '*ondermelk*' (*tr.* undermilk) refers to the location of the more heavy, liquid milk fraction, which becomes concentrated below the solid, floating cream.

The way in which butter is separated from the de-fatted buttermilk varies depending on the method of churning. The butter can be collected by hand or with an implement, such as a spoon. The English term '*buttermilk*' refers to the milk product from which butter is separated, whereas the Dutch term '*karnemelk*' refers to the process of churning (Dutch: *karnen*). Some south-west Asian words for this kind of butter are '*thögh*' (Afghanistan and Tajikistan: Oudenhoven & Haider, 2015); '*dugh*' (western Iran: Mortensen, 1993; Watson, 1979); '*shaneenah*' and '*makheed*' (Turkey, Syria, Iraq, Jordan, and Saudi Arabia: Dagher, 1991); '*laban*' (Qatar: Ferdinand, 1993); and '*laban khad*' (Egypt: Dagher, 1991).

Butter that is obtained after churning is called '*ordinary butter*' in English. Some other words for this kind of butter are '*maska*' (Afghanistan and Tajikistan: Oudenhoven & Haider, 2015); '*kıri*' (western Iran: Mortensen, 1993; Watson, 1979); '*samneh*' (Turkey, Syria, Iraq, Jordan, and Saudi Arabia: Dagher,

1991); and '*zibd*' (Qatar: Ferdinand, 1993). Alternative spellings for '*samneh*' are '*saman*' (Jordan: Weir, 1990) and '*samn*' (Syria: Jabbur, 1995).

Butter from which the excess water has been removed (see section 2.4.5) is called '*clarified butter*' (figure 40). The general term in Arab countries is '*semn*' (or '*samn*'). '*Ghee*' is a term that originates from the Indian subcontinent and refers to a similar but more refined kind of clarified butter (Wilson, 2017). Some other words for this kind of butter are '*rughan*' (Iran: Garthwaite, 2009); '*zirdrughan*' (Afghanistan and Tajikistan: Oudenhoven & Haider, 2015); '*rün*' (western Iran: Watson, 1979); and '*dehen*' (Qatar: Ferdinand, 1993).

The clarified butter can be mixed with plant material to improve taste and colour. Flavouring plants that are mentioned for Jordan include *Ruta chalepensis*, *Matricaria chamomilla*, *Artemisia herba-alba*, and *Trigonella foenum-graecum*. Plants that improve the colour include *Carthamus tinctorius* and *Curcuma longa* (Palmer, 2002).

FIGURE 40: Lumps of clarified butter offered for sale at the bazaar of Ardabil. Each lump is individually wrapped in plastic (Ardabil, Iran; July 2017; RC).

2.4.4 Solidifying of proteins

The coagulation of milk proteins is a chemical process. It can be caused by fermentation and by enzyme activity. Heating may also result in the coagulation of milk proteins, but only when the milk has become acidic prior to this heating. Unfermented milk can be heated without affecting the structure of the milk proteins (McGee, 2004). The coagulation of milk proteins can also be improved by adding specific plants. For example, the Ayr Tuareg from Niger use fragmented *Caralluma mouretii* (Nicolaisen & Nicolaisen, 1997).

Milk proteins can be distinguished into two groups according to their sensitivity to acidity. Casein proteins are sensitive to acid, and coagulation (also termed denaturing) of these proteins occurs if the acidity is lowered. Whey proteins are not sensitive to a change in acidity and therefore do not coagulate. When referring to milk processing, the solid, coagulated casein proteins are designated as '*curds*' and the process of coagulation as '*curdling*'. Curdled milk is a mixture of a solid component (curd) and a liquid component that contains whey proteins, fat, milk sugar, and minerals. If de-fatted milk is used for curdling, the fat content is low, and if fermented milk is used for curdling, the milk sugar content is low and lactic acid is present.

In its strict sense, the word '*curd*' has little practical relevance because it only deals with the casein proteins. If curdling takes place during milk processing, the solid mass that is produced will contain not only casein proteins but also some water with suspended and dissolved nutrients. The amount of non-casein compounds present depends on the composition of the raw milk, the fermentation process, the use of enzymes, and the method of removing water (see section 2.4.5). This gives credence to ethnographic descriptions of milk processing in which the word '*curd*' is used in relation to butter making and the making of solid or dried milk products from yoghurt and from fermented de-fatted milk. Edelberg and Jones (1979), for example, use the word 'curd' for the solid pieces of ordinary butter that are separated from the buttermilk. And Watson (1979) uses the words '*curd mass*' (*shiraz*) to designate the drained buttermilk.

The enzyme '*rennin*' (chymosin), which is present in the stomachs of ruminant mammals, facilitates optimal digestion of the milk they drink. Humans can obtain rennin from, for example, slaughtered calves. If milk is an important product of a herd of sheep or goats, the rennin is usually obtained from male calves, because they do not contribute to milk production. In Yemen, cheese makers collect the stomachs of slaughtered calves and dry them for future use. When milk has to be inoculated with rennin, a piece of dried stomach is wetted and then squeezed several times over the milk. The wet stomach is then dried again for future use (Dagher, 1991). Cheese makers in Nuristan (north-eastern Afghanistan) first dry the fresh, inflated stomach by smoking it. Then they pour some milk through a hole and dry the stomach a second time. They then pulverize the dried stomach, mix it with some whey obtained from cheese making (see section 2.4.4), and stir it in a container for 3 days. The final mixture, being the inner part of the stomach of a young mammal that contains rennin and is called '*rennet*', is then ready for making cheese (Edelberg & Jones, 1979).

The coagulation of milk proteins can be the result of only lactic acid fermentation (acid coagulation), of only enzyme activity (enzymatic coagulation), or of a combination of both (combined acid–enzymatic coagulation). Whereas fermentation can start naturally in raw milk due to the presence of naturally occurring lactic acid bacteria, coagulation by enzyme activity is only possible if enzymes are added to the milk. When attempting to sequence traditional foods with respect to their origin, it makes sense to distinguish solidified milk products that are produced by fermentation from those that are produced by inoculation with enzymes (using rennet).

The designation '*non-rennet cheese*' is used here for solid and dry milk products in which the coagulation of the milk proteins was caused only by lactic acid fermentation and from which the whey was removed. An alternative designation is '*acid-coagulated cheese*'. This group of milk products is also referred to in the literature as '*concentrated fermented milk*' or '*strained fermented milk*' (Tamime et al., 2014). These last two terms designate these milk products by the characteristic process applied to the milk (viz. removal of water and fermentation), but they do not label the milk product with its own name.

Three different kinds of fermented milk products can be used for making non-rennet cheese: yoghurt, skimmed milk, and buttermilk. The three kinds of solid non-rennet cheeses that can be made without heating the fermented milk product are '*kariesh*' (from skimmed milk), '*chakka*' (from buttermilk), and '*labneh*' (from yoghurt; '*yoghurt cheese*'). Further dehydration of these solid non-rennet cheeses is possible, and the resulting products are here all designated with the name '*kashk*' (figures 39 and 43; see also section 3.3.1). Although kariesh (alternative spellings: *karish*[*i*]; *kareish*) most probably was originally only made from skimmed cow's milk, it is now also made from cow's yoghurt and cow's buttermilk, and sometimes rennet is added (Dagher, 1991; El-Gendy, 1983; Elhamid, 2016; Todaro et al., 2013). An alternative name for kariesh is '*cream cheese*'. If fermented milk is also heated, a different name can be used to distinguish that cheese from its counterpart that is produced without heating. For example, non-rennet cheese made from heated yoghurt or buttermilk is called '*alyok*' in the Little Pamir region of north-eastern Afghanistan (Oudenhoven & Haider, 2015).

Rennet cheese can be made from different kinds of milk products: unfermented raw milk, sour milk or yoghurt, and de-fatted milk (skimmed milk and buttermilk) that is made from either unfermented or fermented milk (figure 43). Rennet cheese that is allowed to ripen for some time develops a special flavour, taste, colour, and texture due to the activity of specific bacteria and fungi (moulds). Although rennet cheese made from yoghurt clearly differs from non-rennet cheese made from yoghurt, the nutritional value of these two cheeses may be rather similar. Mixing rennet with skimmed milk or

Process				**Milk product**		
Lactic acid fermentation	***De-fatting*** *Creaming*	***De-fatting*** *Churning*	**Hydrolysis**	***Whey not removed***	***Whey removed***	***In brine***
○				yoghurt	labneh	
○	○			skimmed milk	kariesh	mish
○		○		buttermilk	chakka	
○			○	yoghurt	siahmazgi	
	○		○	skimmed milk	?	
		○	○	buttermilk	?	
○	○		○	skimmed milk	?	
○		○	○	buttermilk	?	
			○	raw milk	takammart	halloum

FIGURE 41: Milk processing options for the production of non-rennet cheese and rennet cheese. Some possibilities are not accompanied by the name of a cheese product because the publications we consulted did not document any such product. Milk processing concerns fermentation, de-fatting, inoculation with rennet, removing whey, and adding brine. Names of the milk products are given before the removal of whey and after the removal of whey (hydrolysis = protein coagulation by the enzyme rennin).

buttermilk will result in a low-fat rennet cheese. Although none of the anthropological studies that were consulted for this publication document examples of rennet cheese made from de-fatted milk, we consider the use of de-fatted milk for making rennet cheese a plausible alternative to converting de-fatted milk into non-rennet cheese (figure 41).

Raw milk is usually heated before it is inoculated with some rennet to kill the bacteria that are present. Examples of rennet cheese made from unfermented raw milk are *'takammart'* and *'gibneh beydah'*. Takammart is a traditional rennet cheese (*'akammar'* means rennet) made by Tuaregs and Arab pastoralists living in the Sahara. Fermentation is prevented by heating the raw milk (Nicolaisen & Nicolaisen, 1997). Gibneh beydah (an Arabic word that translates as white cheese; alternative spellings: *'bayda'* and *'baida'*) is a traditional rennet cheese made in many Arabic countries. Alternative names for this cheese are *'gibneh tariyah'* (an Arabic word that translates as soft cheese) and *'gibneh domiata'* (named after the city and governorate of Damietta, in the Nile Delta of Egypt) (Abou-Donia, 1996). Fermentation is prevented by a combination of heating and mixing with much salt (Dagher, 1991). If this cheese is kept in brine, it is called *'halloum[i]'*, *'msannara'* or *'mushallaleh/mushala[l]'* (Hilali et al., 2011; Wilson, 2017). This pickled rennet cheese has a prolonged shelf-life.

'Siahmazgi' is an example of a cheese in which both acid and enzymatic coagulation take place (figure 42). This rennet cheese is produced in the Gilan

FIGURE 42: Siahmazgi cheese offered for sale in the bazaar of Rasht. The fermentation of the curds takes place in a sheepskin, hence its particular shape (Iran; July 2017; RC).

province of Iran. It is made from sheep's milk or a mixture of sheep's and goat's milk. The fermentation of the curds takes place in a sheepskin (Farahani et al., 2014).

2.4.5 Removal of water

To obtain milk products that can be stored for a long time, most of the water has to be removed. Such dehydrating is necessary for ordinary butter; for cheese; and, unless it is used for mixing with cereals, for yoghurt and de-fatted milk.

Two basic methods of water removal can be distinguished. One method concerns the removal of water with its suspended nutrients (fat and protein) and dissolved nutrients (carbohydrates and minerals) (in figure 39: "removing water and nutrients"). This method results in some loss of nutrients in the milk product from which the water is removed. The extracted water with nutrients is called '*whey*', regardless of the kind of milk product from which is has been extracted.

Removing whey from ordinary butter is necessary to prevent the butter from becoming rancid. Therefore, butter is thoroughly kneaded and washed with water. The kneading results in the aggregation of the fat particles into larger clumps and the removal of much of its water. Washing facilitates the extraction of the dissolved nutrients in particular. Removing the whey from non-rennet cheese is traditionally done by draining, in which the yoghurt or de-fatted milk is put in a cloth bag that can be tied to, for example, a tree. Removing the whey from the more solid rennet cheese can be done by '*pressing*' the cheese in a cloth or container.

The whey is still nutritional because it contains milk sugar that has not been converted to lactic acid; whey proteins; casein proteins that have not been coagulated; water soluble vitamins; and minerals. Whey that is extracted from an unfermented milk product is termed sweet whey, and whey that is extracted from a fermented milk product is termed sour whey.

Although whey is relatively poor in nutrients, it is still used. It can be drunk, fed to animals, used in the preparation of a dish, or used in the processing of other milk products. According to Edelberg and Jones (1979), whey that has been separated from rennet cheese is used for diluting the buttermilk during churning in the Little Pamir region in north-eastern Afghanistan.

The second method of removing water concerns the removal of only water, without a reduction in the nutrients present in the milk product (in figure 39: "removing water"). It is achieved by sun-drying (or air-drying) and/or boiling the milk product. While this kind of water removal does not affect the nutritional value, it does increase the water activity, thus improving the preservational qualities of the milk product.

Because milk proteins (both casein and whey) do not coagulate when the temperature rises, it is possible to thicken milk through boiling without curdling it (McGee, 2004). But boiling requires a lot of fuel, and therefore sun-drying is practiced if possible. If no heating (boiling) is applied, the conversion of a liquid milk product into a dried milk product without loss of its nutrients requires a long drying period, with the risk of spoilage due to fungal growth in the interim. If heating is problematic or impossible, it is more efficient to first remove part of the water with its nutrients, followed by the removal of only water by drying in the sun or wind. This kind of drying is used for the production of kashk from either yoghurt or de-fatted milk.

Boiling can be applied to whey to make it into a solid or dry product. The heating of the whey results in a complex of chemical reactions that involve the bonding between milk sugar and milk proteins. The chemical compounds that are produced result in the browning of the milk. The solid or dry product is called '*qara qurut*'.

3
Examples of traditional foods

3.1 Use of cereals and milk

The identification and classification of traditional foods benefits from an integrated approach in which both crop cultivation and animal husbandry are taken into account. Traditional foods can be made from the crop yield, such as grain, or from milk obtained from domesticated mammals. In addition, traditional foods are made by mixing the two, for example, cereal grain and milk (figure 43). Overviews dealing with milk products sometimes classify the cereal-milk product tarhana as a fermented milk product (e.g. Abdelrahman et al., 2010; Dagher, 1991; Wilson, 2017), but in the classification presented below, a cereal-milk product is considered to be a separate category.

There are no English equivalents for the names of certain food products. In that case, a type word has been chosen to represent the particular food product. In order to avoid confusion, the names that have been chosen for the different milk products clearly differ in the way they are written. However, as a consequence, names from different language families are used. For example, the name '*tarhana*', which is used in Turkey and Greece (with a different spelling: '*trahana*'), is used in this publication for the dried cereal-milk product. And the Persian word '*kashk*' is used for dried fermented milk made without rennet, in accordance with its use today in Iran. The names '*tarhana*' and '*kashk*' emphasize the fundamental difference between the two food products they represent. This difference would not be as clear if, for example, we were to use the word '*kishk*' (a loan-word from the Persian kashk; Aubaile-Sallenave, 1994) for the dried cereal-milk product instead of '*tarhana*', because the word '*kishk*' is so similar to the word '*kashk*'.

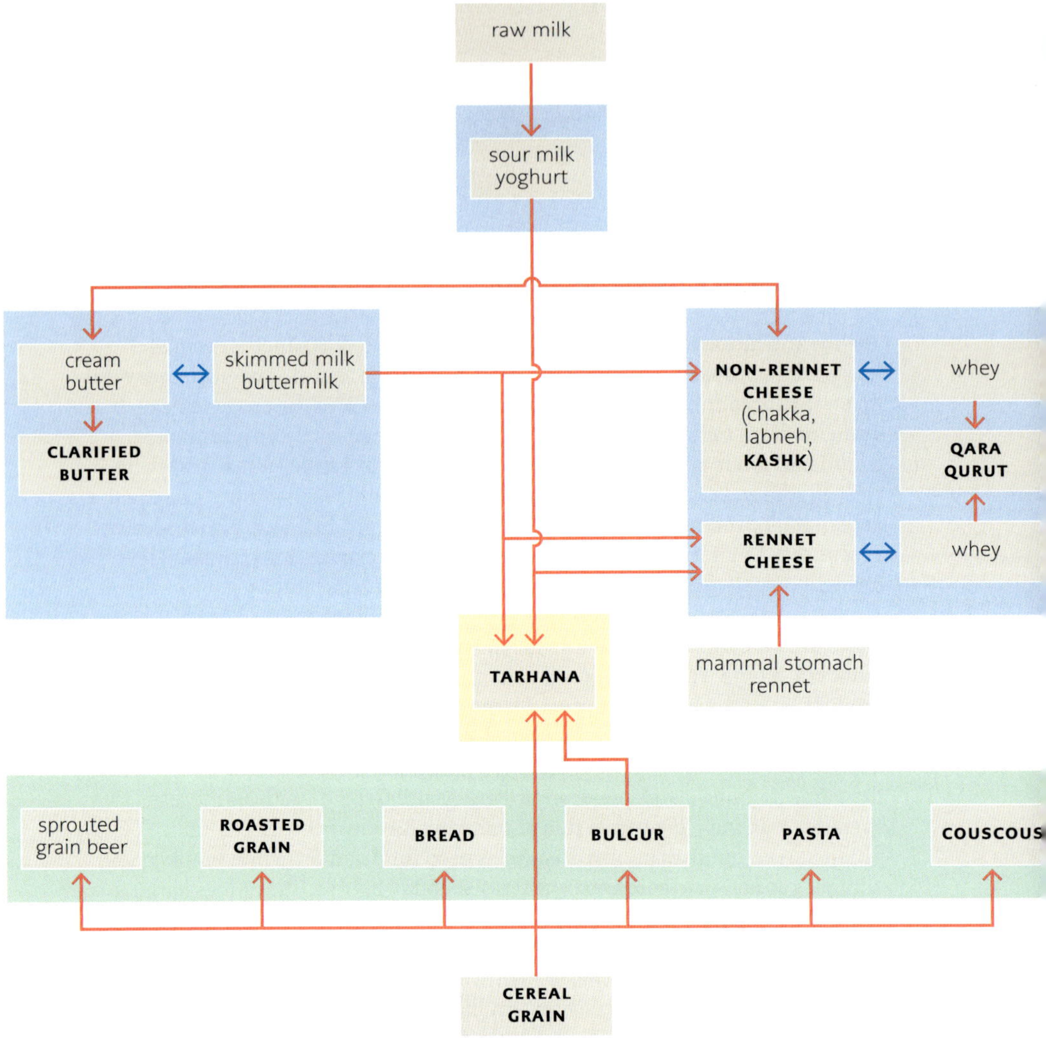

FIGURE 43: Examples of traditional food products. Cereal products are shown in green, milk products are shown in blue, and the cereal–milk product '*tarhana*' is shown in yellow. Single arrows indicate the conversion into a new product, and double arrows indicate the separation into a principal product and a by-product. Food products having a long shelf-life are presented in boldfaced capitals.

3.2 Cereal products

3.2.1 Sprouted grain

Sprouted grain is obtained by adding water to grain kernels. Usually, the grain kernels are first washed and then soaked in fresh water for 2 days. Then the grain is placed in a humid piece of fabric to facilitate further germination. The digestibility of the grain is improved by the enzymatic conversion of starch into sugars.

The sprouted grain is eaten before it turns green. The sprouts can be eaten as a snack or mixed in a salad. A traditional dish of germinated grain and wheat meal is '*samanu*' (alternative spelling: '*samanoo*'), which is made to celebrate the Iranian New Year (the first day of spring). This kind of dish is prepared by means of heating and requires a long period of stirring (Amiryousefi & Razavi, 2013).

3.2.2 Beer

The production of beer includes two successive processes that deal with living organisms: malting, which involves the living grain, and brewing, which involves living yeast. During malting, the cereal grain is allowed to germinate by adding water. Starch-hydrolyzing enzymes are formed during germination, and a small amount of the starch is converted into sugars that can be fermented by yeast. Once a sufficient amount of enzymes have been formed, the sprouted grain (termed *malt*) is killed by heating. Kilns, such as those known from the Roman period in Britain and France, could have been used for this purpose (Van der Veen, 1989). Before the second stage of beer making is initiated, the malt is coarsely ground to make the sugars in the endosperm accessible to the yeast. During the brewing process, the enzymes further convert starch into fermentable sugars. Then alcohol fermentation is performed by adding yeast, in which the fermentable sugars are used as an energy source and converted into ethanol (alcohol) and carbon dioxide (see also sections 2.2.6 and 2.2.7).

Grain provides an optimal return in beer production if its viability and germinability are high. The viability is determined by the percentage of living grain in a sample, whereas the germinability is determined by the speed and uniformity of germination. Grain that is used for beer making therefore needs to meet the same requirements as that which is used for sowing. An optimal return can be achieved by using grain from the most recent harvest.

Other variables that more specifically determine the success of beer making are the chemical composition of the grain, the hull tightness, the kind of yeast that is responsible for the fermentation, the quality of the grain, the amount of starch, and the quality of the water that is used. The amount of starch in the endosperm is of special interest because it is this chemical compound that determines the amount of alcohol that can be made.

A critical stage in beer making is the separation of the liquid from the solid fragments at the final stage of brewing. The liquid consists of water with alcohol, and the solid fragments consist of the coarse cereal fragments and yeast. Hulled cereals are preferred in brewing because their fragments are more solid and can thus be more easily separated from the liquid once fermentation has finished.

The presence of all kinds of impurities, including microorganisms, such as ergot sclerotia, have a negative impact on beer making, and on brewing in particular. This places high demands on the quality of the yield and its storage. Brewers prefer the harvest of an erect and clean crop; they do not like to work with so-called lodged grain, meaning plants that have become bent due to weak culms and strong wind and that thereby become more sensitive to contamination with dirt and all kinds of organisms.

The quality of water is of particular interest during the brewing stage. Fresh and clear water not only optimizes yeast fermentation, but also contributes to the taste of the beer. Beer made from clean water is therefore of a higher quality than beer made from less clean water. Beer made from somewhat brackish water is also of inferior quality.

In south-west Asia, beer is made from different cereals, including Barley (*Hordeum*), Wheat (*Triticum*), Oats (*Avena*), and Rye (*Secale*). Today, special strains of hulled barley are preferred for beer making because of their high starch content. Such desirable strains may already have been available in the remote past thanks to selection by brewers, who can judge the differences in starch content by quantifying the amount of alcohol that is obtained. Using hulled barley, rather than hulled wheat, for beer making also has the advantage that the hull tightness is high. This entails that many of the coarse fragments that are used for the brewing consist of the grain particles that are connected with hull particles. Such solid parts are easily separated from the liquid. The hull in hulled wheat is not tightly connected with the grain, and coarse grinding will result in a mixture of unevenly fragmented chaff (hulls and rachis) and grain. Moreover, the proportion of chaff, which does not contribute to the alcoholic fermentation, is relatively high in hulled wheat. This makes hulled wheat less attractive for beer making from a commercial point of view as well.

Additives are used to improve both the taste and the shelf-life of the beer. Without additives, beer has no distinct taste of its own, in contrast to wine, which is made from grapes or other flavourful fruits. A variety of aromatic plants are used to improve the taste of beer, including Bog myrtle (*Myrica gale*), making the beer taste sweet, and Hop (*Humulus lupulus*), giving the beer a bitter taste (Behre, 1999; Cantrell, 2000). Because the alcohol percentage of beer is much lower than that of wine, the shelf-life of beer is relatively short. These same two plants also improve the shelf-life: Bog myrtle through its bitterness and astringent effect and Hop through its acidity.

3.2.3 Roasted grain

Today, roasted grain is produced in a number of different ways. It can be made from both hulled grain and naked grain and from both partly unripe or ripe grain. The roasting can be applied to the whole grain or, if it has been harvested ripe, also to the polished grain. The grain can be roasted and then granulated or pulverized, or, instead, pulverized and roasted.

In the past, the production of roasted grain differed in relation to the different phases of cereal domestication, as shown schematically in figure 44. Initially, the harvesting of wild cereals was limited to the ripe rachillate triplets (barley) and rachillate spikelets (wheat). Roasted grain could thus only be obtained from ripe grain (figure 44: 1). Once harvesting tools came into use to improve the yield, it became possible to harvest spikes from wild cereals that are partly unripe. Part of the upper ripe rachillate triplets or rachillate spikelets of this harvest can be used for sowing, whereas the remaining part of the spike, consisting of grain that is not yet fully ripe, can be roasted (figure 44: 2).

Only once the cereals had become domesticated and the rachis had become non-brittle (barley) and semi-brittle (wheat) did it become possible to harvest spikes in which all of the grain kernels are fully ripe. If roasted unripe grain is preferred, part of the crop has to be harvested before it is fully ripe (figure 44: 3a). It is also possible to harvest at a later stage and to produce roasted grain from spikes that only contain fully ripe grain kernels (figure 44: 3b).

The availability of naked cereals, being a next step in the domestication process, implies that grain kernels have become the new dispersal unit. Because these cereals have the ability to disperse their grain, the potential yield can be reduced due to the shattering (dispersal) of the grain kernels during harvesting. To prevent this shattering, the crop can be harvested when the spikes are not yet fully ripe. Part of this yield can be processed into roasted unripe grain (figure 44: 4a). Another part should be allowed to after-ripen in order to obtain the quantity of grain that is required for sowing. It is also possible to apply after-ripening to the entire harvest and to use part of this harvest for the production of roasted ripe grain (figure 44: 4b).

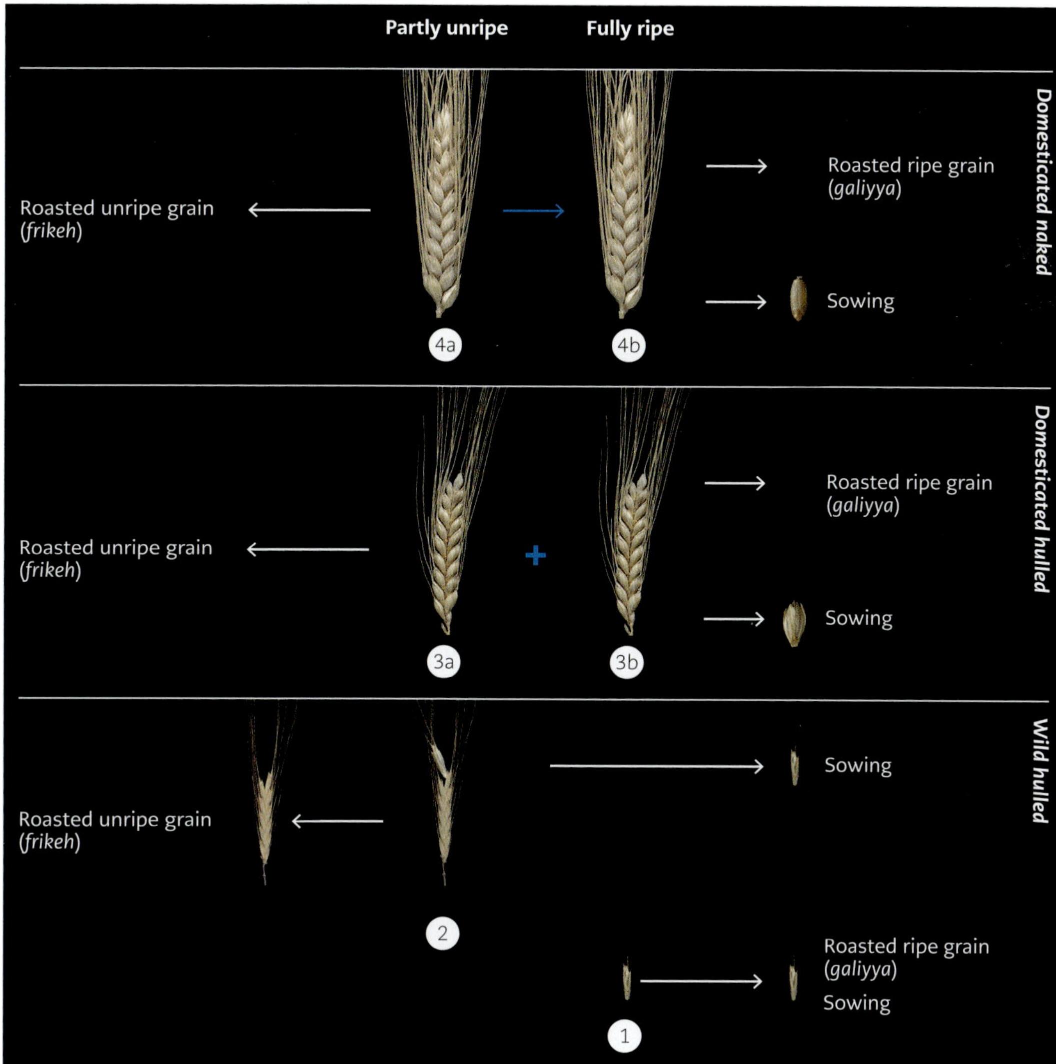

FIGURE 44: Production of roasted grain from wheat at different stages of domestication. The lower half of the figure pertains to wild wheat (1 and 2); the upper half pertains to domesticated wheat (3 and 4). (1): ripe and (2) partly unripe wild wheat; (3a) partly unripe and (3b) ripe domesticated hulled wheat; and (4a) partly unripe and (4b) ripe domesticated naked wheat. Units used for sowing are also indicated (+ means two harvests from the same field [first unripe and later ripe]; ⟶ means a single harvest that is partly allowed to after-ripen).

Here we use the Arabic name *'frikeh'* for unripe roasted grain. Alternative transcriptions from the Arabic include *'fre[e]keh'*, *'fre[e]kah'*, and *'frikkeh'*.

Alternative names for the same food include *'ferike'* (in Kurdish in Iran), *'firig'* (in Turkish), and *'qarakharman'* (in Kurdish in Iraq; in Turkish in Iran) (personal communication, T. Talebi, 2017; Ahmad & Askari, 2015).

Harvesting of the spikes is done either during the first stage of ripening, when the grain is milk-ripe, or during the first part of the second stage of ripening, when the grain is yellow-ripe. During the milk-ripe stage of ripening, chlorophyll is still present in the fruit wall (pericarp), and the grain is soft and milky due to the presence of water in the vacuoles of the endosperm. During the yellow-ripe stage of ripening, the chlorophyll gradually disappears from the fruit wall, the water content further decreases, and the dry-matter content increases. At the end of the yellow-ripening stage, the grain kernel is still soft but no longer milky (Percival, 1921).

Transport of the soluble glucose from the leaves and the culm into the developing grain endosperm continues throughout the ripening process. Once the glucose has arrived in the endosperm, the glucose is converted into starch, which is insoluble and thus does not lower the water activity. As a consequence, at this point, the unripe but full-grown grain has not yet acquired its maximum starch content.

Harvesting the unripe grain should be done at the right time. By crushing the grain between the fingers to squeeze out the milky endosperm, it is possible to judge when the unripe grain has reached the milk-ripe stage (Musselman & Al-Mouslem, 2001). The intensity of the green colour of the hull and grain kernel is another indicator of the ripening stage and the colour can be used to judge when to harvest for a particular desired colour of frikeh. Milk-ripe grain will produce frikeh that is clearly green, whereas early yellow-ripe grain will produce frikeh with only a faintly greenish appearance.

The stages in the production of frikeh are as follows. After harvesting, the grain is first air-dried for some hours. Then it is roasted by setting the harvested plants alight for a short time on a cleaned area outside the field. Al-Azm (2009) mentions some 3–5 minutes for naked wheat. The vegetative parts are burnt and can be used as fertilizer (figure 45). Frikeh can be eaten immediately after roasting (Palmer, 2002). For frikeh not intended for immediate consumption, a second drying stage takes place after roasting; the frikeh grain is air-dried in the shade, resulting in evaporation of water and thereby a reduction in weight of 40 per cent (Musselman & Al-Mouslem, 2001). Quantities that have been stored may require a final heating before consumption.

The harvesting of unripe grain and the production of frikeh are primarily related to yield improvement. Harvesting of unripe grain may also be practiced because food supplies run out before the grain is ripe, or to reduce the risk of crop failure or crop destruction in the intervening period. Once crop production became embedded in an organized food economy, the production of frikeh can also be related to claiming shares and collecting taxes, as it reduces the yield in terms of volume on which taxation must be paid. In this way, the production of frikeh can be indicative of socio-economic conditions (Al-Azm, 2009).

Once frikeh became a regular part of the diet, its production and consumption may have been continued even in the absence of the reasons noted above, because its distinctive taste and flavour are appreciated. This could even result in higher prices for frikeh than for ripe grain. Musselman and Al-Mouslem (2001), for example, mention that fresh undried frikeh was

FIGURE 45: Vegetative plant parts of Hard wheat (*Triticum turgidum* ssp. *durum*) that are produced as a by-product in frikeh preparation. The naked wheat was harvested close to the ground, judging by the number and size of the culm fragments. Only some of the plant remains have become completely charred. The sample was collected from a threshing floor by R. Neef (Marsara, Jordan; 1985; DF).

three times as expensive at the Aleppo market (Syria) than ripe grain, and that fresh dried frikeh was up to eight times as expensive.

Initially, frikeh was only available outside its production area in response to demand from people who had moved from areas where it was part of the traditional cuisine; as such it was primarily offered for sale in shops and markets specialized in non-local food. Today, its widespread availability outside its production area suggests it is popular beyond these groups as well. It is currently being discovered as a potential health food and is now offered for sale also by regular shops. This frikeh can originate from a number of different production areas. In the Netherlands, for example, frikeh can be obtained from production areas in Turkey, Syria, Lebanon, Jordan, and Palestine. This frikeh is made from naked wheat, being either Hard wheat (*Triticum turgidum* ssp. *durum*) or Bread wheat (*Triticum aestivum* ssp. *aestivum*).

The Arabic name '*galiyya*' is used here for ripe roasted grain (Palmer, 2002). In Persian, the name is '*gandomak*'. In Turkey, roasted wheat is designated with the term '*buğday kavurga*', which means 'wheat roasted'. Roasted ripe grain is currently made from bread wheat and is still offered for sale in Turkey and Iran, although not as a regular supply. The grain can be soaked in water or milk and dried again prior to roasting. The roasted grain can be mixed with other roasted seeds or fruits, such as those of Hemp (*Cannabis sativa*) and Pistachio (*Pistacia*). In Ethiopia, roasted Barley (*Hordeum vulgare*) is the main ingredient of a roasted grain mixture known as '*kolo*'. The roasted barley is mixed with roasted seeds of Chick pea (*Cicer arietinum*), Peanut (*Arachis hypogaea*), or Sunflower (*Helianthus annuus*).

The roasting of florets or grain kernels can be done in a concave sack or pan, optionally mixed with sand. The grain can be soaked prior to roasting, through which the grain becomes somewhat swollen. The florets and grain

FIGURE 46: Meal made from Bread wheat (*Triticum aestivum* ssp. *aestivum*), infected by the Confused flour beetle (*Tribolium confusum*). Meal from this same batch was used for bread making in July 2016, and at that point the meal contained only eggs of beetles. The meal shown in the photo had been secured for the reference collection and had been stored at room temperature. One year later, more than 800 adult beetles were present in 1 litre of the meal (obtained in Zaros, Crete; photographed September 2017; DF).

kernels can also be sprinkled with salted water, through which the surface becomes evenly covered with a thin salt spray.

Roasted meal or flour can be made in two different ways: by pulverizing roasted grain or by roasting pulverized grain. The latter method of preparation is mentioned in various classical sources, although it is not always clear which method was applied when texts deal with roasted meal or flour (Peňa-Chocarro & Zapata, 2014).

The reasons for roasting meal or flour are two-fold: roasting improves the shelf-life and roasting facilitates mixing with ingredients that cannot be heated again. Only insects and vermin with special skills can consume whole grain kernels and thus reduce the shelf-life, whereas many more pests can consume pulverized grain and thus spoil the food (figure 46). Roasting makes the grain unpalatable to animals. Roasted grain is therefore avoided by animals and thus can be kept unspoiled for a longer period of time. Meal or flour is also roasted if heating of the mixture in unroasted condition would be problematic, for example in the preparation of kataifi dough, an ingredient for pastries traditionally made in the Mediterranean and west Asia, or *kavut*, a traditional Turkish dish (Karaoğlu & Kotancilar, 2006; figure 47).

Roasted meal or flour is ideal for pastoral nomads because it can be kept for a long period without spoilage. Avitsur (1975) mentions its use by pilgrims who are on the hajj. It can be mixed with a milk product, such as butter or buttermilk, and is then ready to eat. It is also a staple ingredient of nomads living in the Himalaya (including Tibet and Nepal), where it is called *'tsampa'* (alternative spelling: *'tsamba'*) (Jones, 1996). The Berbers in the Maghreb (north-west Africa) separate the roasted barley after grinding it into a coarse

FIGURE 47: Pastry made by mixing kataifi dough and caramelized sugar syrup. The hot mixture is twisted multiple times, until a bundle of papery threads is obtained. This is cut into pieces and decorated with a bow. It is made by a mother and daughter with heat-resistant hands (Kalamafka, Crete, Greece; July 2016; DF).

fraction (*melthuth*) and a fine fraction (*tchicha*), which are stored for winter and spring consumption (Dagher, 1991). In the Canary Islands, the name '*gofio*' is used for roasted meal or flour made, among others, from cereal grain.

3.2.4 Bulgur

Bulgur is a dried '*intermediate food*' made from grain that is first par-boiled and then granulated. Alternative spellings include '*bulgar*', '*bulghur*', and '*burghul*'. In Greece, the same product is called '*pligouri*' (Valamoti, 2011). It is questionable whether granulating has always been a regular part of its production. Boiled whole grain kernels (Arabic: *salīga*) differ from bulgur that it is completely boiled, and the former could have been a by-product of bulgur making (Palmer, 2002). Polishing the grain kernels before granulation is optional in bulgur making.

Par-boiling was originally applied to rachillate spikelets of wheat to improve the subsequent fragmentation and separation into grain kernels and chaff. This process is comparable to the par-boiling of spikelets of Rice (*Oryza sativa*). Assuming that par-boiling was initially only applied to hulled wheat, we can infer that bulgur was originally made from such cereals as Einkorn (*Triticum monococcum* ssp. *monococcum*), Emmer (*Triticum turgidum* ssp. *dicoccon*), and Spelt (*Triticum aestivum* ssp. *spelta*). Although Köksel et al. (1999) mention that in the past also hulled Barley (*Hordeum vulgare*) was used for the production of bulgur, par-boiling is not necessary prior to pearling the barley florets; pearling can be done by pounding wet florets or by grinding dry florets.

Bulgur is still produced from hulled wheat, though on a limited scale. In the Black Sea region in Turkey, bulgur is still made from einkorn. It is available in local shops but is also traded to other regions. Einkorn bulgur was obtained, for example, from a farmers market in Istanbul, where immigrants from the Black Sea area create a demand for this product. There is even an effort underway to achieve a 'slow-food' protected designation for einkorn bulgur, in order to promote its distribution to other areas. Bulgur made from emmer is still produced in western Iran and has been collected by us in East Azerbaijan province (Tabriz), Kurdistan province (Sanandaj), and Ilam province (Ilam).

Today, bulgur is also produced from naked wheat, often Hard wheat (*Triticum turgidum* ssp. *durum*), but also Bread wheat (*Triticum aestivum* ssp. *aestivum*). The par-boiling to separate the grain from its chaff is not necessary in the case of these wheat taxa, so it is merely applied to obtain a similar intermediate food.

Bulgur can be eaten as part of a '*meal*', including as a soup, and needs only a short second heating. The kind of dish in which the bulgur is used is determined by the coarseness of the fragments. The mixture is usually separated in three main qualities: coarse (Turkish: *pilavlik bulgur*), medium (Turkish: *köftelık bulgur*), and fine (Turkish: *çiğköftelık bulgur*). The Turkish names refer to the dishes in which each of them is used. *Pilavlik bulgur* is used in a similar way as rice (Turkish: *pilav*), whereas *köftelık bulgur* is used in making, for example, a spicy mixture of bulgur and lentils (*mercimek köftesi*). *Çiğköftelık bulgur* is used in making a dish that consists of raw meat and spices (*çiğköfte*). Bulgur is also used in the preparation of tarhana. This cereal-milk product also has a long shelf-life and is used for making soup (see section 3.4.1).

3.2.5 Bread

Bread is a baked mixture of pulverized grain, water, and salt. The mixture is called '*batter*' if it is fluid, due to a relatively large amount of water having been used, and '*dough*' if it is solid, due to less water having been used. If bread is to be made from wheat, the chaff has to be separated from the grain, which in hulled wheats is a separate stage in crop processing. When hulled barley is used in bread making, the florets can be pulverized because the hulls (lemma and palea) are tightly connected with the grain kernel and most of it will be pulverized as well. The proportion of hull to grain kernel is small in comparison to wheat, and its presence in food as dietary fibre stimulates the peristaltic movements of the intestine. Optionally, after pulverization of the florets, fine sieving can be used to sift out the coarse fragments of the hull.

Although milk is used as a liquid in products involving cereals, for example, in the cereal-milk product tarhana, in the case of bread, it is questionable whether milk would have been used instead of water in the remote past, for two reasons. The presence of milk lipids could have been problematic because the bread may have become mouldy during the drying process. According to McGee (2004), the whey proteins are also problematic, because they weaken the gluten and result in a dense loaf.

To keep bread for a long time, it must contain very little moisture. This can be accomplished by evaporating the moisture, either during baking or after baking. Only when the dough is flattened to a very thin sheet is it possible for enough moisture to evaporate during the baking process to result in bread that can be kept for a long time. The formation of a crust at the beginning of the baking process is problematic because the crust hampers the evaporation of water. One solution is to use a batter, in a very thin layer. The other is to use dough and to flatten it into a very thin unit. But even in flattened dough, the steam can cause the bread to inflate slightly once the crust has become sufficiently hard. To allow the remaining steam to escape, the flat dough can be perforated before the baking process. This process, called '*docking*', can be done with a simple tool, such as a fragment of a cereal culm. The process can be sped up by performing multiple perforations simultaneously, using a more advanced tool, such as a stamp with a large number of pins. The use of such dockers is often easily discernible from the pattern of the perforation in the baked bread. This pattern also offers the possibility to express the identity of the bread maker.

If the bread still contains too much moisture to allow for long storage after baking, it must dry out after baking. It is important that this drying process does not take too long, mainly in order to prevent the development of mould. The chances of mould forming can be reduced by limiting the percentage of fat. This can be achieved by removing the bran and the germ from the cereal prior to pulverizing. The polishing of grain kernels of wheat and the pearling of florets of barley, both of which lead to removal of the bran and the germ, may have been the result of a desire to produce bread with a long shelf-life.

It is also possible to further dry the bread in the oven using the residual heat of the oven after the first baking process, provided the bread can be placed horizontally. A peel oven meets the requirements for such double baking. A proper insulation of the oven optimizes the drying.

Double baking in an oven can be applied to both thin bread and thick bread. Because a crust is produced in both, the baked bread should be fragmented before it is placed in the oven for the second time. This fragmentation creates fracture surfaces without a crust that then allow the remaining moisture to escape. Flat bread is easily broken or cut into pieces. The fragmentation of thick bread into even pieces can be facilitated by a special folding of the dough or by notching the dough with a knife prior to the first baking (Cappers et al., 2016).

An advantage of large-scale bread production is the efficient use of fuel. If large-scale production takes place at the level of the single household and the aim is to be fuel efficient, it is important that the bread can be stored for a long time. For that to happen, it needs to be able to become sufficiently dry, and it needs to not spoil for any other reasons either. In such a context, leavened bread can be undesirable, especially if the bran and the germ have not been removed, because such bread will spoil more quickly. But if the bread is to be consumed over a short period of time, large-scale production of leavened bread, as happens in bakeries, is no problem.

3.2.6 Pasta

Pasta is a dried intermediate food that consists of pieces of unleavened dough made from a mixture of pulverized grain, water, and salt. The heating is done by boiling. The essential difference with bread is the heating procedure. Bread is dry-heated and then either eaten immediately or stored as an edible food. Most pasta, on the other hand, is originally made without heating and only becomes edible after boiling in water. Partial heating is, however, possible if roasted grain is used. Roasted grain becomes available, for example, when the stubble field is burned after harvesting (figure 48). Pasta made from pulverized roasted grain still needs a final wet-heating to improve its digestibility. Because pasta is wet-heated, its shape differs from that of bread as well. The pasta dough is divided into small, uniform shapes prior to drying to shorten the cooking time.

Although dried pasta made of unroasted grain can be stored before it is heated, it is questionable whether this was done on a large scale in the past. Because heating only takes place prior to consumption, it does not kill eggs of insects that were present at harvest time, and it also does not represent an

FIGURE 48: Roasted grain kernels of Bread wheat (*Triticum aestivum* ssp. *aestivum*) collected from a field after the burning of the stubble. The intensity of roasting varies among the grain kernels (Domuztepe, Turkey; August 2000; DF).

advantageous use of fuel. It seems likely that only pasta made with roasted grain would have benefited from large-scale production and storage, because the roasting kills any eggs and insects and thus extends the shelf-life.

3.2.7 Couscous

Couscous (Berber: *keskesu*) is a dried intermediate food made from a mixture of coarse and fine granulated grain that is mixed with water to cause it to agglomerate into granules of a uniform size. In its basic form, it is made without heating. Both barley and wheat are used for making couscous.

Just like bulgur, couscous initially consists of a mixture of coarse and fine granules. A uniform size is desirable for efficient heating because all particles should need the same cooking time. Whereas in bulgur this is effected by separating the different grades, in couscous this is effected by causing the unequal granules to merge into uniform sized particles. Traditionally, this is achieved by mixing the granulated grain with water, which is a laborious process. It is also possible to roll the fragmented grain into small balls, as described for the Touareg in north-west Africa (Nicolaisen & Nicolaisen, 1997).

Couscous is made without heating. Heating takes place prior to consumption. In order to avoid the disintegration of the aggregated particles during heating, the couscous is typically heated with steam. Two stacked containers are needed for this kind of wet-heating. Water or soup is heated in the bottom container. The couscous is put in the upper container, which has a perforated bottom. This allows the steam from the bottom container to enter the upper container, thereby '*cooking*' the couscous.

Because heating of couscous usually only takes place prior to consumption, it seems unlikely that it was produced and stored on a large scale. In this regard, couscous is similar to pasta.

3.3 Milk products

3.3.1 Kashk

The Persian word '*kashk*' (alternative spelling: *kaslik*) is used in this publication to designate all dried non-rennet cheese. Although in modern-day Iran, kashk unambiguously refers to a dried milk product, the meanings of this Persian word and related words in other language families are rather diverse. Depending on the time period and region, these words may refer to different kinds of milk products, to cereal-based products, or to cereal-milk-based products (Aubaile-Sallenave, 1994). A selection of the words that are used today to designate a dried non-rennet cheese is presented in figure 49.

Word	Region	Reference
awles	Algeria	Nicolaisen & Nicolaisen (1997)
chirpi	Tibet	Dunham (1987)
chura	Tibet	Ekvall (1968); Jones (1996)
gemead	Turkey, Jordan, Syria, Iraq, Saudi Arabia	Dagher (1991)
igt	Saudi Arabia, Kuwait	Ferdinand (1993)
jameed	Jordan, Syria, Iraq	Hilali et al. (2011); Shaker et al. (1999)
kashk	Iran	Aubaile-Sallenave (1994)
keş	Turkey	Personal communication, Y. Okur (2017)
madeer	Saudi Arabia	Dagher (1991)
madir	Saudi Arabia	Dagher (1991)
oggtt	Saudi Arabia	Dagher (1991)
qurut	Afghanistan, Kyrgyzstan	Pedersen (1994); Oudenhoven & Haider (2015)
shankleesh	west Asia, Egypt	Dagher (1991)
tukhp	Afghanistan, Tajikistan	Oudenhoven & Haider (2015)
uggot	Saudi Arabia	Dagher (1991)

FIGURE 49: Selected terms used for a dried non-rennet cheese as documented in different countries and regions.

Kashk is made from dried fermented milk (yoghurt) or from dried fermented de-fatted milk (skimmed milk and buttermilk). Thus, kashk production is either about converting milk (yoghurt) into dry cheese only, or about processing milk into two storable products: butter and dry cheese. If the kashk is made from yoghurt, first the water and nutrients are removed to produce labneh. Then the labneh is further dehydrated by air-drying. The alternative pathway first deals with butter making, either by skimming the floating cream or by churning the sour milk. The de-fatted milk is further processed into kashk by extracting water and nutrients to produce kariesh (from skimmed milk) or chakka (from churned milk) and then by further dehydrating the solid cheese by exposing it to the air (figure 39). Some 2–2.5 kg of kashk can be made out of 40–50 kg (l) of milk.

Creaming easily occurs in cow's milk, whereas it occurs more slowly in milk from sheep, goat, and water buffalo (McGee, 1994). Because the fat particles in cow's milk are light, they easily form a floating cream layer on the surface. This could explain why today, kashk is mostly made from yoghurt and buttermilk rather than from skimmed milk. But the production of the non-rennet cheese kariesh, which is traditionally produced in Arab countries (including Egypt and Sudan), still offers the opportunity to make kashk from skimmed milk. Kariesh is made from water buffalo or cow's milk. Although those in water buffalo milk do not float as readily, the milk's high fat concentration, which can be twice that of goat's milk, facilitates the formation of such cream floating as well.

Both the shape and the composition of kashk are quite variable. Its composition depends on the animal from which the milk is derived and on the specific processing, in which both the method of preparation of butter and the method of water extraction determine the final composition. If milk from different taxa of animals is available, this milk is preferably processed separately in order to profit from its different qualities (Oudenhoven & Haider, 2015). For example, kashk made from cow's milk is lighter and whiter than kashk made from goat's or sheep's milk.

Shankleesh is a dry non-rennet cheese which differs from other types of kashk because fungal growth occurs during ripening. This is possible due to the relatively high moisture content of the cheese. Once the cheese has ripened sufficiently, the outside is cleaned and covered with fragmented Thyme (*Thymus*), which not only improves the taste but also inhibits further fungal growth. Shankleesh could be classified as a mouldy kashk.

Although kashk can be made entirely of milk, some salt is usually added before shaping. Small amounts of starch or flour (1 per cent) can be added to prevent cracking (Hilali et al., 2011). Recently, it has been suggested to mix bran with non-rennet cheese to improve the yield, the moisture content, and the dietary fibre content (Elhamid, 2016). Such use of a high-quality by-product may also have occurred in the past.

The shaping of kashk can be done in several ways. Irregular pieces are obtained by drying the kashk in large lumps and then breaking these into pieces. This kind of shaping is, for example, documented by Oudenhoven and Haider (2015) for the Little Pamir in Afghanistan and by Ferdinand (1993) for Qatar. Traditionally produced Kashk that has been collected by us in western Iran had been shaped by hand into balls or elongated pieces, with the shape of the inner part of the hand visible. A rather quick shaping method involves remodelling the solid cheese into a large, triangular mass, from which small triangles are cut with a thread. Natural imprints of matting may be present on the bottom face of the kashk. Such matting could be put on elevated constructions (Persian: *talvareh*) to prevent animals from touching the kashk (Huang, 2009). Possibly as a reminder of this matting, artificial decorations are sometimes made into the shaped units of this cheese before drying. Ekvall (1968) mentions a kind of kashk that is made into hard, round cakes with a hole in the middle so that the dried pieces can be strung.

Kashk has excellent preservation qualities and can be stored in a fabric bag or open container for two years. But a smaller quantity is usually made so that the maximum storage time does not exceed one year, because new kashk can be made once the milking season starts up again. Its long storage life is possible due to its high acidity, high salt concentration, and low moisture content. Salt is often partly concentrated on the outside.

Kashk can be eaten as a snack or used in the preparation of a dish. It is made into a soup, which can be eaten with rice and meat, or it is pulverized by pounding and mixed with some water to obtain a drink (Hermanns, 1949; Shaker et al., 1999). Watson (1979) mentions that in western Iran kashk is also used as fodder for sheep and goats during winter time.

3.3.2 Qara qurut

'*Qara qurut*' is solid or dry whey. The milk product is referred to as '*whey cheese*', '*black cheese*', or '*black kashk*' (Persian: *kashk siyah*). Based on its method of preparation and composition, it is justified to classify qara qurut as a milk product with its own identity, equal to butter and cheese (figure 43). The Turkish name '*qara qurut*' is closely related to '*chu ra*', which refers to dry buttermilk (*chakka*) in Tibet.

Whey is a by-product of cheese making. Although whey becomes available in the production of both non-rennet cheese and rennet cheese, today, production only takes place with whey obtained from the processing of non-rennet cheese, being chakka and labneh (figure 39).

Qara qurut is obtained by heating the whey so that the water evaporates. During the final stage of the heating, caramelization and maillard reactions take place, which result in chemical conversion of any milk sugar and water-soluble whey proteins still present. This changes the colour of the concentrated qara qurut, making its tint darken as the temperature increases. When sour milk is used, the qara qurut contains lactic acid. Light brown qara qurut, which is also referred to as '*white qara qurut*', is less concentrated than dark brown qara qurut, and the former quality can therefore be used in slightly larger portions. Dark brown qara qurut, on the other hand, can be extremely sour tasting.

Qara qurut contains few nutrients in comparison with butter and cheese, and its production requires a lot of fuel because a relatively large amount of water has to evaporate. This could explain why the production of qara qurut is limited. At present, it is still produced in Tibet, Afghanistan, and Iran. In Iran, qara qurut can be acquired in markets in Urmia (West Azerbaijan province), Ardabil (Ardabil province), and Isfahan (Isfahan province). In Isfahan, balls are offered for sale of which one half consists of white kashk and the other half of dark brown qara qurut. This combination thus combines the two dry milk products made from labneh or chakka.

Although qara qurut is primarily a food product, it is also used in Tibet for tanning leather (Ekvall, 1968). And Goldstein and Beall (1990) mention the use of qara qurut as a cosmetic product in Tibet. Called '*döja*', it is put on the forehead, nose, and cheeks by women between the ages of 10 and 40.

3.4 Cereal-milk products

3.4.1 Tarhana

Tarhana is a dry fermented cereal-milk product. Other, optional ingredients may include salt, yeast, vegetables, herbs, spices, and fruits. Alternative spellings include '*tarkhineh*' (Iran) and '*trachana*[s]', '*trachano*[s]', '*trakahana*[s]', and '*tarhonya*' (Greece). The same product is called '*xinochondros*' on the island of Crete and '*chackla*' on the island of Lesbos (Valamoti, 2011), whereas it is called '*kushuk*' in Iraq and '*kishk*' or '*khisk*' in Jordan, Lebanon, Syria, and Egypt.

Tarhana is quite a variable food in both shape and composition, and hence in nutritional value. This is due to the variation in the type of grain and the type of milk product used, and to the variation in the ratio in which these basic ingredients are mixed, which varies from 2 to 4 units of grain to 1 unit of milk (Tamime & O'Connor, 1995). The use of additional ingredients is optional. The mode of preparation varies with respect to fragmentation, heating, fermentation, and shaping. Nowadays, barley is not used in tarhana production. But experimental production has shown that barley flour, obtained from both hulled and naked forms, can be used instead of wheat or in combination with wheat (Erkal et al., 2006).

In south-west Asia, the production of tarhana takes place in summer, when the grain harvest is available, domestic animals give milk, and solar heat allows for rapid drying. The final moisture content of tarhana is 6–12 (–20) per cent (Herken & Çon, 2014; Tamer et al., 2007).

The use of whole grain kernels in tarhana production has been reported by Steinkraus (1983). Fragmentation of the grain is optional. It may involve polishing, granulation, or pulverization of the grain kernels. The tarhana samples recently collected by us in Turkey, Iran, Egypt, and Greece are all made from granulated or pulverized grain. In Jordan, it is the finest fraction of bulgur that is used for mixing with the milk product (Palmer, 2002).

Tarhana can be made without heating. In that case, the grain is soaked or pulverized to meal or flour, so that only a relatively short cooking time is needed during the final food preparation. The unheated and pulverized grain used to make tarhana is referred to as '*chondros*' in Greece (Valamoti, 2011).

Heating can be applied to only the milk, to only the grain, or to the cereal-milk mixture. The heating of only the raw milk is aimed at killing the microorganisms present, but does not affect the final cooking time required for the final dish. The heating of the grain most probably has its origin in the use of hulled wheat, because, unlike grain kernels of hulled barley, grain kernels of hulled wheat must be separated from the vegetative parts of the rachillate spikelet (see section 2.3.3). Par-boiling of the rachillate spikelets precedes the fragmentation, and the grain thus becomes available as bulgur (Cappers et al., 2016). The current use of hulled wheat in the production of tarhana is recorded for Eskişehir and Afyon, in Turkey (Tamer et al., 2007). When naked grain is used in the production of tarhana, as is the case today, the grain is usually mixed with the milk and then heated for some time.

Two types of fermentation can play a role in the preparation of tarhana: lactic acid fermentation and alcoholic fermentation (Herken & Çon, 2014; Kivanç & Funda, 2017). Lactic acid fermentation takes place in the milk. Buttermilk from fermented milk or yoghurt is often used for the preparation of tarhana. Tarhana made of fermented milk or yoghurt is designated as '*sour tarhana*', whereas tarhana made of unfermented milk is designated as '*sweet tarhana*'. Tarhana usually has a pH value of 3.5–5.0. The use of de-fatted milk rather than yoghurt has the advantage that this milk contains little fat and therefore is less susceptible to the development of mould while drying. If necessary, yeast is added to the cereal-milk mixture, which is responsible for alcoholic fermentation. Fermentation takes place after mixing and during the first stage of the drying process, and it can vary from 1 day to 1 week.

Prior to the final drying process, the tarhana is shaped into small units that easily dry and that can be easily dosed into the final dish. These units consist of uniform granules, compact pieces or flakes. Granular tarhana is obtained by rubbing the dried tarhana through a sieve in order to obtain granules of a uniform size. The compact pieces of tarhana collected by us from north-western Iran have a length or diameter of (4–) 8 (–20) cm (sd: 2.6), a thickness of (0.7–) 1.8 (–6.0) cm (sd: 1.8), and a weight of (6–) 39 (–225) g (sd: 33). Such pieces can be shaped by hand or in a mould. The imprint of the palm and/or fingers shows how the tarhana was formed. Today, the use of moulds seems to be limited. A special example, however, are the relatively large tarhana pieces made by the Luri people of western Iran, which are made using both hand-shaping and moulds. The tarhana is shaped like a barley floret, with a clear ventral groove on one side (See Atlas section). An impression of the structure of the substrate on which the tarhana was dried can be visible on the underside.

Vegetables, herbs, and spices are added to the tarhana to improve its flavour, taste, and colour. Sporadically, other ingredients are also added, such as grape (*Vitis vinifera*) and Cornelian cherry (*Cornus mas*). Tarhana that contains grape has been collected by us in Sanandaj (Iranian Kurdistan province).

By drying the tarhana in the shade, the colour of added vegetables is retained as much as possible. Vegetables that are widely used include Turnip (*Brassica rapa*), Onion (*Allium cepa*), Tomatoes (*Solanum lycopersicum*), and Red pepper (*Capsicum annuum*). Herbs and spices commonly used today include Coriander (*Coriandrum sativum*), Dragon sagewort (*Artemisia dracunculus*), Peppermint (*Mentha × piperita*), Parsley (*Petroselinum crispum*), Dill (*Anethum graveolens*), Thyme (*Thymus* spp.), and some members of the genus *Echinophora* (Apiaceae). The use of Turmeric (*Curcuma longa*) makes the tarhana bright yellow, but the colour disappears after a few months. The colour intensity is thus indicative of how long ago the tarhana was produced.

Some members of the genus *Echinophora* (Apiaceae) have a tradition of use as a flavouring agent in milk products, including tarhana. *Echinophora tenuifolia* ssp. *sibthorpiana*, whose distribution area includes parts of Turkey, Greece, and Iran, has even acquired the name '*tarhana herb*'. Mixing ground Tarhana herb with other tarhana ingredients results in an increase in the populations of both lactic acid bacteria and yeast, but the impact on the acidity of the dough seems to be negligible (Değirmencioğlu et al., 2005). Most probably, its use is related to the presence of an essential oil that improves the flavour and could also protect the stored tarhana against vermin (Abbas et al., 2015).

Large plant parts are chopped prior to mixing. In the Kurdistan province of Iran, the tarhana that includes Turnip is referred to as '*tarhana shalmineh*' ('*šhalam*' being a variety of turnip [*Brassica rapa*] with elongated and sour leaves), in contrast to the regular tarhana, which is referred to as '*tarhana dooineh*' ('*dooineh*' being dough).

Contamination may occur when the tarhana is spread out for drying. Dusty impurities, such as soil particles and hairs, can become attached to the surface. Unchopped, broad leaves are used as a kind of coaster during the drying process. Large fragments of such leaves remain stuck to the underside when the tarhana is collected after drying. Once the tarhana is dried, it has no hygroscopic properties and can be stored for several years in an open container (Dagher, 1991; Değirmencioğlu et al., 2005). Its acidity, salinity, and low moisture content, as well as the possible antimicrobial properties of some additives, also contribute to its long shelf-life.

Tarhana is edible after fermentation. It is common practice to use it for soup making. But it can also be eaten as a chip, which is nowadays especially done in the Kahramanmaraş region of Turkey (Özçam et al., 2014).

4
Sequencing traditional foods

4.1 Processing cereals and milk

The question is whether the various forms of traditional foods made from barley, wheat, and milk that are still made today were also made by the late hunter-gatherers and the first farmers—and whether it is possible to sequence these forms of traditional foods from basic to more advanced products. In order to sequence traditional food products in this sense, it will be necessary to take into account the availability of the ingredients, the desirability of preserving the food, and the technological developments and material culture related to these foods. In the case of cereals, the harvesting phase (unripe or ripe) and the morphology of the harvesting units (more specifically, the proportion of cereal grain to vegetative parts) are also relevant.

The differing natures of grain and milk each determine the development of food preparation in their own way (figure 50). Processing the harvest of grain is necessary to make it edible and digestible. But the shelf-life of the grain, whether or not supported by vegetative plant parts, can be rather long, provided sufficient care is taken with their storage. For example, in the agadirs of Morocco—the well-protected food storage facilities in which people could also find temporary safety—grain has been stored for up to 25 years (Grammet, 1998). Large-scale production and storage of grain products can be triggered by the need to make efficient use of fuel. An additional advantage of heating the harvest in an early stage of crop processing is the killing of eggs and pests present in the grain stock.

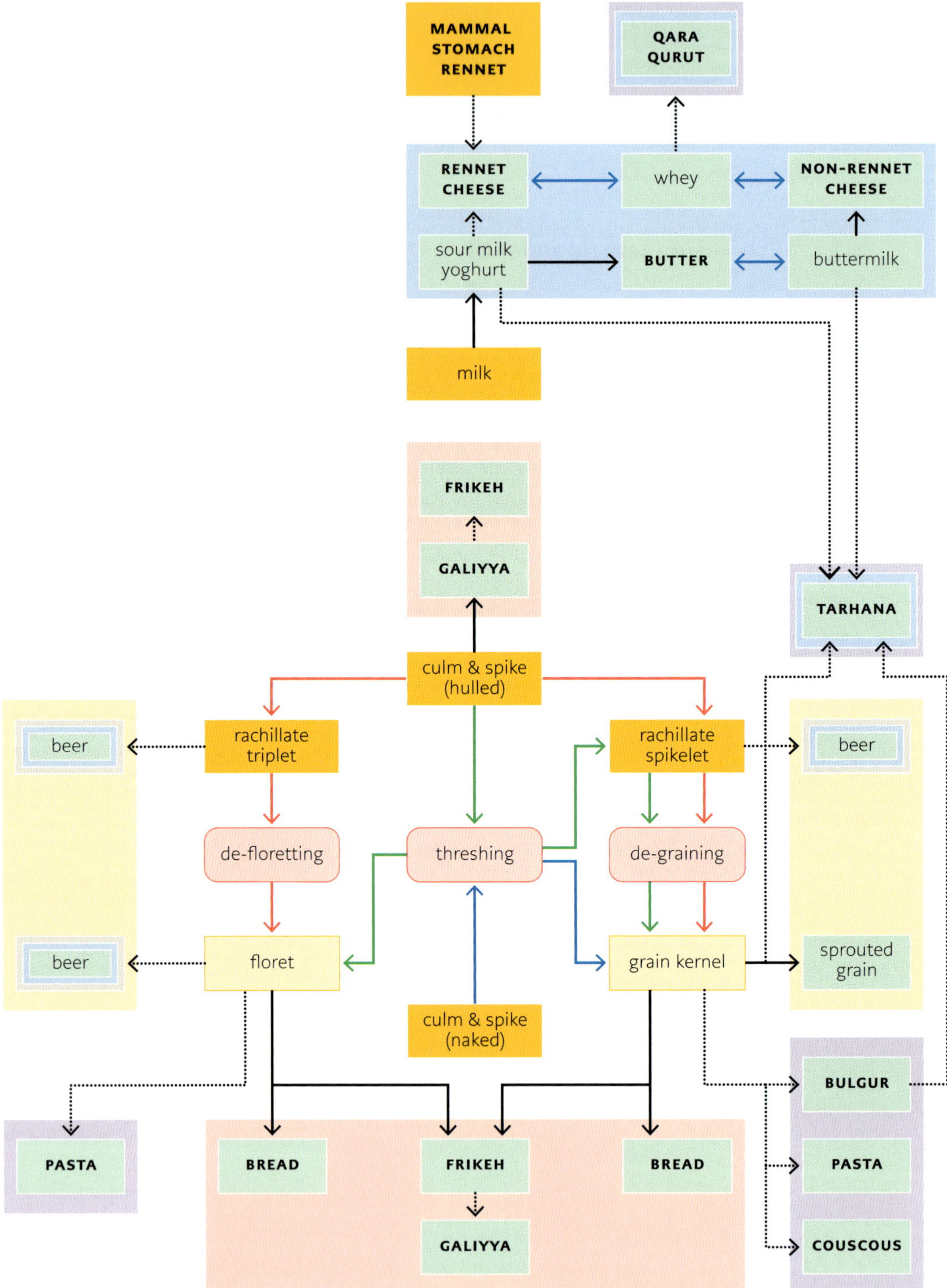

FIGURE 50: Production of cereal products, milk products, and cereal–milk products from rachillate triplets, rachillate spikelets, florets, grain kernels, milk, and rennet. The harvesting units of barley and wheat, milk, and rennet are shown in bright yellow, and

the post-harvesting units are shown in light yellow. Three fragmentation processes are presented: de-floretting, de-graining, and threshing. The combination of grain ripeness, rachis brittleness, and hull tightness is indicated with coloured arrows (red arrow = harvesting ripe rachillate triplets/spikelets from wild cereals; green arrow = harvesting partly unripe spikes from wild cereals and harvesting fully ripe spikes from domesticated hulled cereals; blue arrow = harvesting partly unripe or fully ripe spikes from naked cereals). Black solid arrows indicate food processing that is considered to be of primary origin, and black dotted arrows indicate food processing that is considered to be of secondary origin. Food products with a limited shelf-life are in lowercase and those with a long shelf-life are in capitals. The colours of the borders surrounding the food products indicate fermentation (blue), germination (brown), dry-heating (pink), and wet-heating (purple). Fragmentation of harvesting units and threshing units is not indicated.

Unlike grain, milk can be consumed without special treatment, but its shelf-life in untreated form is limited. Therefore, the conversion of milk into products that can be kept for a long time becomes necessary once a surplus is produced. Because lactic acid fermentation is a basic process that easily occurs, people ensure that milk sugar is converted into lactic acid so that its shelf-life is prolonged. Additionally, after this conversion, the digestibility of milk products is no longer problematic.

4.2 Sprouted grain

Grain can be made edible and digestible by germinating or by heating. Sprouted grain that is still covered by their hulls is only suitable for beer production. Therefore the consumption of grain sprouts most probably became an attractive option only once naked grain became available, some 10,000 years ago. It is striking that the consumption of grain sprouts today is mainly related to celebrating new life, and thus is more related to ceremonial meals, whereas concentrations of sprouted grain found in archaeological contexts are usually related to beer production.

4.3 Roasted grain

There are several ways of heating grain, involving either dry-heating or wet-heating. The simplest form of heating grain is to roast it, which is a form of dry-heating. When hulled grain is roasted, the grain kernels themselves can then simply be separated from the chaff by rubbing. This form of food preparation is suitable for both wild cereals and domesticated hulled cereals. Because the first stage of harvesting by hunter-gatherers deals only with the ripe grain from the upper part of the spikes, it is likely that roasted ripe grain (Arabic: *galiyya*) is the oldest edible form of heated grain (figure 50). The production of roasted unripe grain (Arabic: *frikeh*) will only have started once whole spikes of wild cereals started to be harvested at the start of the ripening process in order to optimize the harvest. But even at a later stage of cereal domestication, when naked grains became available, the harvesting of unripe grain will have been desirable, but now to minimize crop loss.

The roasting of grain is possible without additional fuel if the plant is cut off close to the ground. The dry, vegetative parts provide the fuel with which the grains are roasted. This harvesting method is especially functional for wild cereals and domesticated hulled cereals. The roasting can take place on a flat, preferably clean surface.

A more advanced and thus more recent form of dry-heating by means of roasting concerns the dry-heating of only grain kernels. This kind of roasting requires an additional kind of fuel, as well as a container to prevent the grain kernels from rolling away. A disadvantage of such pan-roasting is that heat transfer from the container is not evenly spread over the surface of the grain kernel, and continuous stirring is necessary. To reduce the chances of scorching, the roasting of the grain kernels can be done with the aid of sand. For this form of pan-roasting, a high-quality sieve is needed with which the sand can be completely separated from the roasted grain when roasting is finished (Cappers et al., 2016).

4.4 Bread

A still more advanced form of dry-heating involves baking batter or dough. The small grain kernels are merged into a larger unit that is more easily manipulated during the heating process. Dry-heating on a flat stone or in soil becomes easier. To obtain batter or dough, the grain must be granulated or pulverized in order to be able to mix it with water or milk. For this kind of fragmentation, the same implements can be used that are already available for the fragmentation of the rachillate triplets of barley and the rachillate spikelets of wheat. The later development of bread moulds and ovens makes it possible to efficiently isolate batter and dough from fuel, ash, and soil. This will have had three major advantages. The first is to limit the damage to the teeth of the consumer. The second is to increase the shelf-life. And the third, which is related to the second, is to make efficient use of fuel.

4.5 Partially prepared foods

In contrast to dry-heating, wet-heating has the advantage that the heating process is easier to manipulate. It is technologically more demanding than dry-heating because it requires a container that is fire-resistant. It is therefore plausible that food that is prepared by wet-heating was developed at a later stage than food that is prepared by dry-heating. Examples of foods that are prepared by wet-heating are pasta, bulgur, couscous, and tarhana. Although organic materials, such as leather and wood, can be used for wet-heating, the development of pottery and the production of metal containers will have given this form of food preparation an important stimulus. Pottery is suitable for the storage of substantial quantities of cereal yields and of dried food products made from cereals and milk. It also improved the preparation and the consumption of meals. Large scale wet-heating, however, will only have become possible with the development of metal technology. Such a connection between change in staple foods and the spread of iron cooking pots is also mentioned by Leach (1990).

Wet-heating can be done by boiling or steaming. The preparation of pasta and bulgur involves boiling, whereas that of couscous involves steaming. Pasta is preferably made from pulverized grain mixed with water. Pasta dough is prepared in small units and can thus easily be dosed in quantities that are then wet-heated in relatively small containers made from organic material or in ceramic vessels. The availability of pottery may have facilitated the production of pasta with respect to both mixing and wet-heating. The production of bulgur seems to be related to the par-boiling of rachillate spikelets of wheat, which can be obtained from both wild and domesticated forms. This makes it likely that bulgur was originally only made from hulled wheat; current bulgur production from naked wheat should be seen as the production of a similar food product that is much appreciated in traditional cuisine. Par-boiling of rachillate spikelets is preferably done on a large scale, and this sets different requirements for the containers used for heating compared with pasta. Most probably, the large-scale par-boiling only became a realistic option once large metal containers became available. It is suggested by us that the attractiveness of cultivating hulled wheat taxa increased from the Chalcolithic period onwards. Testing this hypothesis would require (1) modelling crop selection in a way that takes into account crop proportions and (2) archaeological evidence for large metal containers. Couscous is prepared by steaming, a method that can be regarded as more technologically advanced and thus less ancient than the methods used to produce pasta and bulgur. For this kind of wet-heating, two stacked containers are required. Water or soup is boiled in the bottom container, and the couscous is steamed in the upper container, which has a perforated base. Such perforated containers can be made from earthenware, but also from wood or basketry (Nicolaisen & Nicolaisen, 1997).

4.6 Beer

Beer could have been made at an early stage of the exploitation of barley and wheat. After all, beer is preferably made from hulled cereals, which means that the rachillate triplets of barley and rachillate spikelets of wheat collected in the wild can be used for beer production. Nevertheless, the most suitable starting material for beer making is the floret of barley, because it contains relatively little hulls and the hulls are also strongly fused with the grain, so after starch hydrolysis the sweet wort can be easily separated from the solid fraction. This would have made beer production an attractive option once barley became domesticated, because it is the fertile floret that becomes separated from the spike by threshing.

Beer making requires both living grain and yeast and includes a number of different phases, with malting, fragmenting, and brewing being the basic steps. Especially when beer is made on a large scale and demands are made on quality, the process requires both the necessary technology and suitable equipment. The malting needs to be done in shallow layers, in which mats or shallow ceramics can be used (Samuel, 2000). It should also be realized that for a maximum yield of beer, it is desirable to use florets from the new harvest because their germination capacity is optimal and beer production is thus maximal. Such use of a new harvest is, however, only justified if the harvest is substantial or if sufficient stock from earlier harvests is available for other purposes. After all, grain that is stored for some years can still be used for the preparation of food in which germination plays no role. These considerations will affect decision making around when to use the harvest.

4.7 Milk products

It is still unclear when people began to milk their domestic animals for the first time. Remains of milk have been found on pottery dating back to 6300 BCE (Evershed et al., 2008). It cannot be ruled out that milking started earlier and that the initial phase of milking took place even before the development of earthenware in south-east Asia, dating back to 7000 BCE. The possible production of a milk surplus in the remote past can be deduced from the ratio of the human population to the animal population and from the sex ratio and the age composition of the latter.
Milking results in a behavioural adaptation in the animals, whereby they have to be milked in addition to being suckled by their own young. The processes that are necessary for the preservation of milk are fairly basic. The clumping of fat particles in milk takes place naturally. The biochemical processes that play a role in converting milk into products with a longer shelf-life also take place under natural conditions. The material culture needed to optimize the preparation of butter and cheese is simple and can be made entirely from organic materials, such as wood, leather, and wool. Moreover, heating is not necessary in the preparation of butter and cheese.

It is therefore possible that the production of butter and cheese took place fairly soon after mammals first started to be milked. The combination of butter and cheese production is plausible because both products are produced in a fixed sequence, each with their specific nutritional composition. And cheese made from fat-free milk is easily sun-dried during the hot summer period, which partly coincides with the lactation period. The conversion of the whey, being a by-product of cheese making, into a product with a long shelf-life, such as qara qurut, probably constitutes a later stage in milk processing and may only have been applied on a limited scale because it requires a relatively large amount of fuel.

The production of rennet cheese will also have been developed at a later stage, because its invention was precipitated by milk coming into contact with the inside of one of the stomachs of neonates, in which the enzyme rennin is present in large quantities. This innovation is most probably related to the selective slaughter of young male calves in order to optimize the milk production of the herd, because the stomach that produces the enzyme rennin has its largest volume in the first weeks after birth, and the use of such stomachs for milk storage will have resulted in a rapid solidifying of milk proteins.

4.8 Tarhana

The production of tarhana only becomes possible once both grain and milk are available. The desirability of mixing grain and milk into one durable product seems to be related to the par-boiling of hulled wheat. A large-scale production of tarhana requires good coordination of the pre-processing of wheat and milk, as well as large (metal) containers for the parboiling of the rachillate spikelets. The current use of bulgur made from naked wheat in the production of tarhana has to be considered a more recent development.

5
Food and cultural identity

Food plays an important role in daily life, both when it comes to regular meals and when it comes to meals served on special occasions. It is therefore not surprising that food contributes to cultural identity.

The cultural identity of our food can manifest itself at both the micro level and the macro level. At the micro level, it concerns, for example, the preparation of a food product according to an old family recipe, or the use of private starters for making sourdough bread or yogurt, or decorating the food product with personal patterns (figure 51).

FIGURE 51: A small quantity of sourdough from the previous baking session is submerged in oil. It is used as a starter and will be mixed with a large quantity of new meal (a small amount has already been shaken into the mixture). Such starters can be quite old and are often used by a single family. Because of the unique composition of fermenting microorganisms, the bread will have its own characteristic taste and structure (west of Katharon, Crete; July 2016; RC).

At a macro level, cultural identity relates to a region and to a considerable time depth. Both the ingredients and the related preparation methods determine the cultural identity. Once trade routes come into being, ingredients become part of the commodities being traded and will have an impact on the cuisine in remote areas. These adaptations concern not only condiments, such as herbs and spices, but also food products that form the basis of a meal, such as cereal grains and tubers.

When we focus on south-west Asia, as the region where agriculture first developed, arable farming was initially based on a combination of three grains (providing carbohydrates), four legumes (providing proteins), and flax (providing both vegetable oil and textile fibres). The first domesticated cereals were Barley (*Hordeum vulgare*) and two wheat species, namely, Einkorn (*Triticum monococcum* ssp. *monococcum*) and Emmer (*Triticum turgidum* ssp. *dicoccon*). The first domesticated legumes were Lentil (*Lens culinaris*), Pea (*Pisum sativum*), Chick pea (*Cicer arietinum*), and Bitter vetch (*Vicia ervilia*). Livestock provided both milk and meat. It is this group of founder crops, milk, and meat that characterizes the early cuisine of south-west Asia. This early cuisine clearly differs from that of other regions where agriculture developed independently, as shown by the comprehensive list of characteristic edible plants for each region where agriculture developed independently presented by Harlan (1992). For example, the American cuisine is characterized by, among other foods, Maize (*Zea mays*); Kidney bean (*Phaseolus vulgaris*); and various species in the nightshade family (Solanaceae), including the Potato (*Solanum tuberosum*), Tomato (*Solanum lycopersicum*), and all kinds of peppers (*Capsicum annuum*). Milk, on the other hand, plays no role in early agriculture in the Americas.

Only a limited number of traditional food items and dishes are still restricted to only those ingredients from plants and animals that were locally available at the time when agriculture first developed. Their basic composition is primary related to their efficient production, in which labour and fuel are the main determining factors. Food that requires a minimum of input of labour and fuel will give a maximum rate of return on energy. Especially when large-scale food production is desired, such food items are much appreciated.

Traditional foods may have a rather pronounced taste due to their basic ingredients and their preservation methods. People usually appreciate the taste of dishes from home more than variations from elsewhere because people learn to eat these ingredients at a very early age and get used to them. However, the fact that, once exotic ingredients become available, people gradually incorporate these exotic ingredients into their traditional cuisine suggest that the taste of food is judged not just by its familiarity, but also by other factors. In this way, exotic products can even supplant traditional products. If a traditional dish goes out of use, it may even happen that local people who were never exposed to it before do not like it. This is, for example, the case with *kishik* (tarhana) in the Gaza area, where children growing up in refugee camps and towns have never eaten this traditional dish. When they have the opportunity to taste it, they judge it to be unpalatable (El-Haddad & Schmitt, 2011).

The appreciation of certain foods may also relate to the methods used in food preservation. Food preservation is mainly based on reducing the acidity and/or on lowering the water activity by removing water and by adding much honey, sugar, or salt. This implies that when food preservation by such traditional methods is practised, people learn to eat food with a pronounced acidic, sweet, or salty taste. It also implies that the incorporation of new food items in the traditional cuisine is more readily accepted when their taste spectrum coincides with that of the cuisine's traditional foods.

The preparation of traditional foods is still practiced. And even when they represent a basic cuisine from a remote past, they still have the potential of contributing to our current cuisine. Their basic composition offers the opportunity of adding other ingredients in order to prepare a more varied cuisine. This can be done during the mixing stage, as is done, for example, in the preparation of tarhana. Both mixing and shaping offer the opportunity to produce a wide variety of tarhana types, and such a traditional food might easily contribute to the cultural identity. Although exotic ingredients, such as Tomato (*Solanum lycopersicum*) and Red pepper (*Capsicum annuum*), are nowadays often added, it is also possible to choose spices, herbs, and vegetable seasonings that are of local origin.

Certain food products of a country have already become part of the Intangible Cultural Heritage of the UNESCO, such as *keskek* of the Turkish cuisine, and the aim in Turkey is to obtain a similar status for the bulgur made from Einkorn (*Triticum monococcum* ssp. *monococcum*), one of the founder crops of agriculture of south-west Asia (figure 52). This shows that food products whose tradition is based on more than 10 000 harvests and a multitude of food preparations still have a promising future.

FIGURE 52: This ekşili pilav is made from coarsely ground Einkorn (*Triticum monococcum* ssp. *monococcum*) bulgur and jam made from apples (*Malus sylvestris*) (Kastamonu, Turkey; August 2015; RC).

ATLAS

The atlas section presents examples of technology that can be used in traditional crop processing and food preparation, as well as examples of traditional foods that are made from grain and/or milk. The technology and each of the food types are briefly introduced at the start of the relevant sections. With respect to the various food items, the characteristics of their composition and processing are summarized and the possibility of early production is discussed.

Following these brief introductions are entries for individual accessions of traditional foods that were collected during recent fieldwork and that now form part of the reference collection at the Groningen Institute of Archaeology (GIA). Each description of an individual accession of a traditional food item starts with information on its origin, whereby a distinction is made between the location where the food item was obtained and the location where is was produced (where applicable). The latter location is mentioned when the food item has been traded. We emphasize that the sampling of traditional foods was not done systematically and that therefore the geographical region that is recorded as the area of origin may represent only part of the area(s) where the food in question is still being produced and traded. Size and weight are mentioned when the food concerns individual units.

The second part of the description concerns the composition of the food item at the level of the individual ingredients and any impurities present. Both of these descriptions are qualitative rather than quantitative. The description of the composition is based on information that was submitted by the person from whom the food item was obtained, on information available on the package (where applicable), and on microscopic analysis. This implies that the descriptions are not always complete, especially with respect to the milk component. The description of the impurities is based on microscopic analysis of the entire sample, which varies in weight. The presence of impurities, which is typical for some of the modern samples of frikeh and bulgur, is indicative of the quality of the food product. However, it should be noted that such food containing impurities is eventually carefully cleaned by hand-picking. This is usually done as part of food preparation, to the small quantity that is to be processed into a final dish.

The last part of the description deals with the morphology of the food. Where possible, the morphology is related back to the processes that have been applied and the implements and installations that have been used in its production. If mentioned on the package, the nutritional value per 100 g is presented as well. Following each description is a photo.

6
Technology

When we assume that a certain food item could have been produced by late hunter-gatherers or by early farmers we are -implicitly- also assuming that the technology that is needed for its production was available.

In order to gain more insight into the early development of crop processing and food production, the study of technology and material culture by different disciplines within archaeology needs to become more integrated with the study of subfossil plant and animal remains.

To this end, the first section of this atlas presents some examples of the different basic processes that are used in the crop processing and food preparation of traditional foods made from grain and milk. These examples of processes and related material culture are followed by some photos, taken over the past 25 years, showing people engaged in traditional food preparation.

Fragmentation by pounding

FIGURE 53: Storage of wooden pestles and a wooden mortar in a barn. Two sizes of pestles are present. At the time this photo was taken, they had been in use for some 7 years (Thiès, Senegal; October 2016; RC).

FIGURE 54: A stone mortar at a communal spot in a village. Such mortars were in use until recently for polishing grain. The beating is done with wooden hammers. Stone mortars can still be found in situ, whereas the hammers have become rare objects. The hammers are usually stored in a barn. Here, they are exposed in the mortar prior to the polishing of grain (Göcer, Turkey; July 2013; YO).

FIGURE 55: Women threshing whole spikes of Pearl millet (*Pennisetum glaucum*) by using a wooden mortar and wooden pestles (Ndofongon, Senegal; October 2016; RC).

Fragmentation by grinding

FIGURE 56: A large quernstone located at the edge of a village. Most probably, each user kept her own handstone at home (Sarab Karzan, Iran; April 2017; RC).

FIGURE 57: Grinding a mixture of grains. The grinding stone is placed on the inside of a sheepskin. Some grain is poured from the operator's hand into the central hole. The handle is turned in alternating directions, with the left and right hands. The handle is improvised, and adapted to the stone for this demonstration. The grinding is accompanied with a specific song (Skoura, Morocco; March 2016; RC).

Separation by winnowing

FIGURE 58: Winnowing by using a sieve. In this way, impurities are removed that are equal to or larger in size than the grain, but lighter in weight. The sieve is composed of wood and the skin of an animal, which is made into thin strips that are woven into a mesh (Çakırbağ, Turkey; July 2013; YO).

Separation by sieving

FIGURE 59: Cleaning fragmented bulgur by using a coarse and a fine sieve. The sieves are made of wood and animal skin. The final check is performed by hand picking (Ihsangazi, Turkey; August 2015; RC).

FIGURE 60: Sieving wheat meal prior to mixing. Large particles, including fragments of the hulls, remain on the fine screen, whereas the purified meal passes through the screen. This sieve is made of wood and a metal screen. Long animal hairs, such as those from a horse's tail, are also used for the screen (Kalamafka, Crete; June 2016; RC).

Mixing

FIGURE 61: Dough is made by mixing wheat flour, dry yeast, milk powder, salt, and clean warm water. Quantities are not measured but are determined by experience (Abu Bakra, Iraq; June 2014; RC).

FIGURE 62: Mixing sourdough with some flour and oil. This mixture will be allowed to rise for some hours and is then mixed with a large quantity of flour and some salt. The mixing is done with the fists (Kalamafka, Crete; June 2016; RC).

Shaping

FIGURE 63: A wooden bowl used for the processing of milk. According to Edelberg & Jones (1979), this kind of bowl is used in particular for pressing the whey out of the cheese, and it gives the cheese its final shape. This bowl was made from a single piece of wood by using an adze. A piece of textile is attached to the handle to make it easier to hang up the bowl when it is not in use. The bowl was collected by N. van Oudenhoven in the 1970s and dates back to the early 20th century. Height: 22 cm; diameter: 29 cm (provenance Nuristan, Afghanistan; 1972–1975; DF).

FIGURE 64: Storage of wooden rolling pins in a barn. The rolling pins are used for shaping the dough balls into paper-thin flaps that are be baked on a convex metal plate. Some rolling pins are fresh cut from branches and are still greenish (Çağlayan, Turkey; May 2015; RC).

Heating

FIGURE 65: A simple heating installation at a picnic site, consisting of three stones that can support a metal plate. Only the metal plate has to be carried in transit (vicinity of Skoura, Morocco; March 2016; RC).

FIGURE 66: Heating for bread baking in an above-ground tannur, a type of pre-heated oven. The inner part consists of earthenware, and it has an opening in the rim through which fuel is added prior to baking. The outside has been built up with tempered clay, and the oven is located in a dedicated room to retain as much heat as possible (Skoura, Morocco; March 2016; RC).

Cooling and drying

FIGURE 67: Scored loaves of bread that have been baked in a peel oven. They have to cool down before they can be fragmented into slices and dried using the residual heat of the oven .

FIGURE 68: Pieces have been broken off scored, baked loaves of bread and put back in the oven for a second time to dry. The oven will be closed to retain its residual heat, and the slices will become sufficiently dry for long-term storage (figures 67 and 68: west of Katharon, Crete; July 2016; RC).

Fermentation

FIGURE 69: After the first, bulk fermentation, the dough is divided into individual loaves. Each loaf is then incised crosswise with a knife, at narrow intervals. Then a second fermentation takes place, which is called 'proving'. The loaves increase in size, and the small incisions widen. The incisions facilitate the fragmentation after the first baking so that the slices can be easily dehydrated in the oven and then stored for a long time (west of Katharon, Crete; July 2016; RC).

FIGURE 70: Proving of loaves of bread in the sun. Each loaf is supported by a platter made of clay, and most of them are placed on a construction made of leaf rachises of the Date palm (Phoenix dactylifera). The bread is called sun bread, a name that recalls this traditional kind of fermentation (Quseir, Egypt; November 1996; RC).

Making frikeh

FIGURES 71 and 72: Modern processing of Hard wheat (*Tritium turgidum* ssp. *durum*) into frikeh. Because only spikes are processed, which reduces the amount of vegetative plant parts available to be used as fuel, Butane gas is used as an additional fuel. First the spikes are sun-dried, then they are concentrated in heaps with a pitchfork and set on fire (top). The roasted spikes are threshed and separated into roasted grain and burnt chaff by means of a combine (Ajaa, Palestine; May 2016; HM).

Making bulgur

FIGURE 73: Par-boiling of a large quantity of rachillate spikelets of Einkorn (*Tritium monococcum* ssp. *monococcum*). Traditionally, a large-sized container called a kazan is used. Here, half an oil drum is used instead. The flat bottom is supported by a metal frame. It is assumed that such large-scale crop-food processing only became possible once metalworking had evolved.

FIGURE 74: Fragmentation of the par-boiled grain of einkorn by means of two large grinding stones. Small quantities of chaff are still present in the grain (figures 73 and 74: Ihsangazi, Turkey; August 2015; RC).

Making bread

FIGURES 75 and 76: Baking flat breads in pre-heated soil. The meal, water, and salt are mixed in a tub. The flattened dough placed on the hot soil and covered with hot sand and ash for some 15 minutes. Then, the ash and soil are moved aside, the bread is flipped on its axis, and it is covered again with the sand and ash for another 15 minutes. To prevent dental damage, it is important to create a smooth surface before baking so that after baking the bread can be easily and carefully freed of soil particles before consumption. Especially stones are problematic in this respect. The bread in the upper photo seems to be of a better quality (Berenike, Egypt; February 1998; RC).

FIGURE 77: Baking bread in a direct-heated, single-chambered peel oven made of tempered clay. The oven floor is covered with large pebbles, making it easy to separate out the baked bread. The dough is shaped on the head of the peel. Imprints of the hand (upper face) and of the pebbles (lower face) in the bread are characteristic of bread baked in this kind of oven (Marrakesh, Morocco; March 2016; RC).

FIGURE 78: Dough is stuck to the inner upright face of a tannur using both hands. The inside of the tannur consists of thin pottery; the outside of this pottery is covered with tempered clay. The bread will retain the somewhat curved shape of the wall of the tanur after baking. The oven is built in a small building to optimize the heating of the oven. A disadvantage is the concentration of the smoke, especially in the upper part of the small room (Skoura, Morocco; March 2016; RC).

FIGURE 79: Baking flat bread at a campsite. At all stages, the food is isolated from the soil, fuel, and ash. The dough is prepared and allowed to rise in metal bowls. Shaping is done with a rolling pin over a low wooden table. The convex metal plate ('saç') is supported by three stones and heated with some pieces of wood. Mixing, shaping, and baking are done by three daughters. They are 12, 15, and 17 years old and started making bread at the age of 10. Some 100 breads will be made today, sufficient to feed a family consisting of two adults and six children for five days (Çatören, Turkey, May 2015; RC).

Making pasta

FIGURE 80: The dough is shaped into a long cylinder by rolling it with the fingers of two hands. Next, the cylinder is divided into small units (Kalamafka, Crete; July 2016; RC).

FIGURE 81: Several pieces of dough are rolled out and cut into square units. Then each unit is rolled up around a metal pin to obtain a tubular piece of pasta (Kritsa, Crete; July 2016; RC).

Making couscous

FIGURES 82 and 83: A couscous of Fonio (Digitaria exilis) is prepared by steaming. In the bottom pan, water is boiling. The steam enters the upper pan through its perforated bottom. The steam ensures that the small, agglomerated granules do not fall apart and are evenly cooked (Colibantan, Senegal; October 2016; RC).

Making butter and buttermilk

FIGURE 84: A transportable installation, made of wood, rope, and the skin of goat or sheep, is filled with milk and expanded by adding air. The milk is separated into butterfat and buttermilk by shaking, a process called 'churning'. The butterfat adheres to the skin, after which the buttermilk can be drained off (Tunis, Egypt; October 2005; RC).

FIGURE 85: Churning raw milk in the courtyard. The skin is suspended from a wooden tripod by means of two ropes. The metal bowl below the churning skin contains butter that has already been separated. The goat is muzzled and her teats are filled with milk (Ezbet Dawud, Egypt; November 2010; RC).

FIGURE 86: Milk processing in a nomad family's so-called black tent. The tent-cloth is made of black goat hair, hence its name. The walls consist of rush mats interwoven with dark brown yarn. The milk is separated into butterfat and buttermilk by shaking it in a metal container. The churning process is performed inside a tent (southeast of Jafar Adab, Iran; April 2017; RC).

Making tarhana

FIGURE 87: Mixing the polished and coarsely ground bulgur of Hard wheat (*Triticum turgidum* ssp. *durum*) with yoghurt and salt. The metal pot stands on an improvised wooden table.

FIGURE 88: Shaping the tarhana using both hands. A shallow impression of a fingertip is made in the middle. The shaped tarhana is collected on a plate and will be sun-dried (figures 87 and 88: southwest of Yılanğömü, Turkey; August 2013; YO).

7
Sprouted grain

Sprouted grain is grain that has germinated. The only example that is depicted concerns germinated barley that is produced for beer production. No evidence of consumption of such malt has been found. It is possible, however, that this kind of processed grain has also been eaten as such and possibly also dried for storage.

Sprouted grain

Obtained	Bolsward (province of Friesland, the Netherlands)
Origin	–
Date	March 2014
Collector(s)	R. T. J. Cappers
Context	Brewery
GIA-accession	34724
Length/Diameter	–
Thickness	–
Weight	–
Quantity	–
Photographer	D. Fennema

Composition

Hordeum vulgare ssp. *distichon*	Hulled 2-row Barley	floret

Description

The germinated grain has been sampled during the malting process. The grain kernels are covered by their hulls, and only the roots are visible. The elongated shoot is still covered by the lemma at this stage of germination. The germinated grain originates from Brewery Us Heit BV and is preserved in alcohol.

FIGURE 89

delveseh.com

8
Galiyya

Galiyya is an Arabic name for roasted ripe grain kernels. Galiyya is considered to be one of the most traditional examples of a cereal food because the first stage of harvesting wild cereals was limited to only ripe grain and because the processing of grain into galiyya requires only dry-heating, which can be done without special equipment. The roasted ripe grain is not suitable for grinding, not even coarse grinding.

Originally, dry-heating had a twofold purpose in cereal processing: facilitating the removal of the chaff and making the grain more digestible. Once barley became domesticated, the removal of the rachis and spiny awns and glumes was easily achieved after threshing by winnowing and/or sieving, and the removal of the two remaining hulls that still tightly cover the grain kernel was not necessary. In domesticated hulled barley, therefore, the dry-heating is primarily applied to improve the digestibility. This is supported by modern samples from Ethiopia, which concern roasted florets of barley (that is, grain kernels with the hulls still present). The transition from wild to domesticated hulled wheat, in contrast, did not change the morphology of the harvesting unit, and the removal of the chaff remained necessary. Therefore, it is plausible that dry-heating was applied to hulled domesticated wheat as one of the methods to facilitate the removal of the chaff. Unfortunately, all of the modern samples of roasted wheat that we have collected concern naked wheat, where the roasting has only been applied to improve its digestion.

Before roasting, grain kernels can be polished or soaked for some time. The roasted florets and grain can be mixed with other roasted seeds or fruits. Today, roasted ripe florets and grain are still offered for sale in shops and markets, but on a limited scale.

Galiyya

Obtained	Groningen (province of Groningen, the Netherlands)
Origin	China
Date	November 2015
Collector(s)	R. T. J. Cappers
Context	Shop
GIA-accession	39658
Length/Diameter	–
Thickness	–
Weight	–
Quantity	–
Photographer	D. Fennema

Composition

Hordeum vulgare ssp. *distichon*	Hulled 2-row Barley	floret

Impurities		
Galium	Bedstraw	fruit (mericarp)
Hordeum vulgare ssp. *distichon*	Hulled 2-row Barley	floret (burnt)
Hordeum vulgare ssp. *distichon*	Hulled 2-row Barley	rachis fragments
Hordeum vulgare ssp. *vulgare*	Hulled 6-row Barley	floret
Hordeum vulgare ssp. *vulgare*	Hulled 6-row Barley	rachis fragments
Persicaria	Knotweed	fruit
Setaria	Bristle grass	spikelet
–	lumps of clay	

Description

The grain kernels have been roasted in the chaff (florets). The degree of puffing varies from none to very pronounced. Scorch marks are present, though only to a small degree. Some florets still retain their original shape, whereas others have cracked open, exposing the white endosperm. The cracks are mostly on the lateral faces, where lemma and palea are connected. A few fragments of the rachis of 2-row barley are present, as well as a few fragments of the rachis of 6-row barley. Although 2-row barley is part of the intended ingredients, both sets of rachis fragments are considered to be an impurity. The rachis fragments are partly still connected to the florets and may bear one or both sterile lateral spikelets (and, in the case of 6-row barley, a fertile floret). Other impurities concern burnt diaspores of arable weeds and clumps of clay. Some lemmas also still bear part of the awn. The sample was bought at a supermarket specializing in oriental foods. The label on the package mentions roasted wheat.

FIGURE 90

Galiyya

Obtained	Amsterdam (province of Noord Holland, the Netherlands)
Origin	Ethiopia
Date	December 2017
Collector(s)	R. T. J. Cappers & Y. Okur
Context	Shop
GIA-accession	41098
Length/Diameter	–
Thickness	–
Weight	–
Quantity	–
Photographer	D. Fennema

Composition

Hordeum vulgare	Hulled Barley	grain kernel
Cicer arietinum	Chick pea	seed
–	salt	

Description

This roasted grain is mixed with a small quantity of chick pea. This combination represents a cereal and a pulse from among the founder crops of south-west Asia. Only the chaff (lemma and palea) has been removed from the barley florets. Remains of the chaff (palea) are especially visible in the hilum groove. The grain kernels have become swollen and the bran has broken lengthwise, whereby the starchy endosperm has become visible. The seeds of the chick pea are unprocessed. The local word for this product is 'kolo', which translates as 'roasted barley'. This product is still a traditional snack in Ethiopia. This sample was bought in an Ethiopian shop in Amsterdam.

FIGURE 91

Galiyya

Obtained	Amsterdam (province of Noord Holland, the Netherlands)
Origin	Ethiopia
Date	March 2017
Collector(s)	R. T. J. Cappers & Y. Okur
Context	Shop
GIA-accession	40330
Length/Diameter	–
Thickness	–
Weight	–
Quantity	–
Photographer	D. Fennema

Composition

Arachis hypogaea	Peanut	seed
Hordeum vulgare	Hulled Barley	grain kernel
–	salt	

Description

The roasted grain is mixed with a small quantity of peanut. Only the chaff (lemma and palea) has been removed from the barley florets. Remains of the chaff (palea) are especially visible in the hilum groove. The grain kernels have become swollen and the bran has broken lengthwise, whereby the starchy endosperm has become visible. The seeds of the peanut are polished and split. The local word for this product is '*kolo*', which translates as 'roasted barley'. This product is still a traditional snack in Ethiopia. This sample was bought in an Ethiopian shop in Amsterdam.

FIGURE 92

Galiyya

Obtained	Sivas (Sivas province, Turkey)
Origin	–
Date	February 2017
Collector(s)	E. Boybeyi
Context	–
GIA-accession	40296
Length/Diameter	–
Thickness	–
Weight	–
Quantity	–
Photographer	D. Fennema

Composition

Secale cereale	Rye	grain kernel
Triticum aestivum ssp. *aestivum*	Bread wheat	grain kernel

Impurities

Centaurea cyanus	Cornflower	fruit
Elytrigia repens	Common couch	floret
Rumex	Docks	fruit with perianth
Triticum aestivum ssp. *aestivum*	Bread wheat	chaff
–	stone	

Description

Home-made roasted grain made of unpolished grain kernels. The grain consists of a mixture of bread wheat and rye, the latter in small quantities. Judging by the glossy surface, it seems probable that fat was used in the heating process. The intensity of roasting varies among the grain kernels, as does the presence of scorch marks. Chaff remains of bread wheat concern glumes and fragments of the awn.

FIGURE 93

Galiyya

Obtained	Van (Van province, Turkey)
Origin	–
Date	2015
Collector(s)	H. Balcı
Context	–
GIA-accession	40025
Length/Diameter	–
Thickness	–
Weight	–
Quantity	–
Photographer	D. Fennema

Composition

Triticum aestivum ssp. *aestivum*	Bread wheat	grain kernel

Impurities

cf. *Lathyrus*	Pea	seed

Description

Home-made roasted grain made of unpolished grain kernels. The grain kernels have been roasted in animal fat, giving the surface a somewhat glossy appearance. Usually sheep's fat, taken from the waist area, is used for this purpose. It is unclear if the grain kernels were soaked before roasting.

FIGURE 94

Galiyya

Obtained	Göcer (Karaman province, Turkey)
Origin	–
Date	May 2015
Collector(s)	Y. Okur
Context	Private house
GIA-accession	38317
Length/Diameter	–
Thickness	–
Weight	–
Quantity	–
Photographer	D. Fennema

Composition

Triticum aestivum ssp. *aestivum*	Bread wheat	grain kernel
–	salt	

Impurities		
Avena fatua	Wild oat	floret
Secale cereale	Rye	grain kernel
Triticum aestivum ssp. *aestivum*	Bread wheat	burnt grain

Description

Home-made roasted grain made of unpolished grain kernels. The grain kernels have been sprinkled with salted water prior to roasting. Small scorch marks are present. The impurities consist of a few burnt grain kernels, a grain kernel of rye, and a diaspore of an arable weed.

FIGURE 95

Galiyya

Obtained	Karaman (Karaman province, Turkey)
Origin	–
Date	September 2014
Collector(s)	Y. Okur
Context	Market
GIA-accession	35103
Length/Diameter	–
Thickness	–
Weight	–
Quantity	–
Photographer	D. Fennema

Composition

Triticum aestivum ssp. *aestivum*	Bread wheat	grain kernel

Impurities

Cannabis sativa	Hemp	fruit
Hordeum vulgare	Hulled Barley	floret
Secale cereale	Rye	grain kernel
Triticum aestivum ssp. *aestivum*	Bread wheat	burnt grain

Description

Home-made roasted grain that was offered for sale at a local market. Unpolished grain kernels have been used for making the roasted grain. The name in Turkish for this product is '*kavrulmuş bugday*', which translates as '*roasted grain*'. Judging from the glossy surface, it seems likely that fat was used in the heating process. Most of the grain kernels have small scorch marks on the fruit wall. The weakest point of the bran is the hilum groove and the lateral part of the grain kernel. It is in these regions that the bran bursts open and the starchy endosperm becomes visible. A small quantity of burnt grain kernels is present. These burnt specimens are a bit smaller than the roasted and puffed grain kernels. Although in some types of galiyya fruits of hemp are also used for roasting and mixed with roasted wheat, the small quantity of hemp in this sample is considered to be an unintended admixture.

FIGURE 95

Galiyya

Obtained	Karaman (Karaman province, Turkey)
Origin	–
Date	August 2014
Collector(s)	Y. Okur
Context	Market
GIA-accession	34964
Length/Diameter	–
Thickness	–
Weight	–
Quantity	–
Photographer	D. Fennema

Composition

Triticum aestivum ssp. *aestivum*	Bread wheat	grain kernel

Impurities

Triticum aestivum ssp. *aestivum*	Bread wheat	grain kernel

Description

Home-made roasted grain that was offered for sale at a local market. The unpolished grain kernels have been roasted. Only small scorch marks are present. The bran may have become isolated from the starchy endosperm at these spots, resulting in puffed scorch marks. Only a fragment of a burnt grain kernel was detected in the sample. The uniform quality of the roasted grain and the almost absence of impurities indicates that this home-made roasted grain was thoroughly checked by hand picking before it was offered for sale.

FIGURE 96

Galiyya

Obtained	Karaman (Karaman province, Turkey)
Origin	–
Date	September 2014
Collector(s)	Y. Okur
Context	Market
GIA-accession	35103
Length/Diameter	–
Thickness	–
Weight	–
Quantity	–
Photographer	D. Fennema

Composition

Triticum aestivum ssp. *aestivum*	Bread wheat	grain kernel
Impurities		
Cannabis sativa	Hemp	fruit
Hordeum vulgare	Hulled Barley	floret
Secale cereale	Rye	grain kernel
Triticum aestivum ssp. *aestivum*	Bread wheat	burnt grain

Description

Home-made roasted grain that was offered for sale at a local market. Unpolished grain kernels have been used for making the roasted grain. The name in Turkish for this product is '*kavrulmuş bugday*', which translates as '*roasted grain*'. Judging from the glossy surface, it seems likely that fat was used in the heating process. Most of the grain kernels have small scorch marks on the fruit wall. The weakest point of the bran is the hilum groove and the lateral part of the grain kernel. It is in these regions that the bran bursts open and the starchy endosperm becomes visible. A small quantity of burnt grain kernels is present. These burnt specimens are a bit smaller than the roasted and puffed grain kernels. Although in some types of galiyya fruits of hemp are also used for roasting and mixed with roasted wheat, the small quantity of hemp in this sample is considered to be an unintended admixture.

FIGURE 97

Galiyya

Obtained	Karaman (Karaman province, Turkey)
Origin	–
Date	August 2014
Collector(s)	Y. Okur
Context	Market
GIA-accession	34963
Length/Diameter	–
Thickness	–
Weight	–
Quantity	–
Photographer	D. Fennema

Composition

Triticum aestivum ssp. *aestivum*	Bread wheat	grain kernel

Impurities

Hordeum vulgare	Hulled Barley	floret
Triticum aestivum ssp. *aestivum*	Bread wheat	burnt grain

Description

Home-made roasted grain that was offered for sale at a local market. The grain kernels have been polished and coarsely ground before roasting. The absence of the bran has resulted in a somewhat tuberculate surface. Remains of the bran are only present in the hilum groove. Scorch marks are present but rare. Impurities are present but rare and include fragments of burnt grain kernels and a fragment of a pearled barley grain kernel.

FIGURE 98

Galiyya

Obtained	Rasht (Gilan province, Iran)
Origin	possibly Qazvin (Qazvin province, Iran)
Date	July 2017
Collector(s)	R. T. J. Cappers & M. Kaffashi
Context	Bazaar
GIA-accession	40881
Length/Diameter	–
Thickness	–
Weight	–
Quantity	–
Photographer	D. Fennema

Composition

Triticum aestivum ssp. *aestivum*	Bread wheat	grain kernel
Impurities		
Aegilops tauschii	Goat beard	rachillate spikelet
Fabaceae spp.	Pea family	seed
Galium	Bedstraws	fruit (mericarp)
Hordeum vulgare ssp. *vulgare*	Hulled 6-row Barley	floret
Ranunculus arvensis	Corn buttercup	fruit
Secale cereale	Rye	grain kernel
Triticum aestivum ssp. *aestivum*	Bread wheat	culm

Description

The grain kernels have been soaked in water overnight, then dried, and then roasted. After roasting, the grain was sprayed with salt water. The roasting time was rather limited and the grain kernels are still quite hard. But this galiyya is considered good quality. The roasted grain is not a common supply in the bazaar. The impurities consist of a mixture of arable weeds and two other cereal crops: rye and hulled 6-row barley.

FIGURE 99

Galiyya

Obtained	Isfahan (Isfahan province, Iran)
Origin	–
Date	April 2017
Collector(s)	R.T.J. Cappers & M. Kaffashi
Context	Bazaar
GIA-accession	40438
Length/Diameter	–
Thickness	–
Weight	–
Quantity	–
Photographer	D. Fennema

Composition

Triticum aestivum ssp. *aestivum*	Bread wheat	grain kernel

Impurities		
Apiaceae	Carrot family	fruit (mericarp)
Convolvulus arvensis	Field bindweed	seed
Hordeum vulgare ssp. *vulgare*	Hulled 6-row Barley	floret
Secale cereale	Rye	grain kernel
Triticum aestivum ssp. *aestivum*	Bread wheat	burnt grain
Triticum aestivum ssp. *aestivum*	Bread wheat	culm fragment
Triticum aestivum ssp. *aestivum*	Bread wheat	culm-rachis fragment

Description

The grain kernels have been soaked for a short period. This results in the uptake of water, through which the grain kernel becomes swollen and the fruit wall that covers the embryo ruptures. In a small proportion of the grain kernels, the embryo has even become visible. The grains were also sprayed with salt water after roasting. The predominant impurities concern grain kernels of rye, florets of barley, and culm fragments of bread wheat.

FIGURE 100

Galiyya

Obtained	Tabriz (East Azerbaijan province, Iran)
Origin	Argentina
Date	July 2017
Collector(s)	R. T. J. Cappers & M. Kaffashi
Context	Shop
GIA-accession	40917
Length/Diameter	–
Thickness	–
Weight	–
Quantity	–
Photographer	D. Fennema

Composition

Triticum aestivum ssp. *aestivum*	Bread wheat	Grain kernel
–	salt	

Description

The roasted whole grain kernels have become puffed. The grain kernels have increased in size and the fruit wall has broken, whereby the spongy, starchy endosperm has become exposed.

FIGURE 101

Galiyya

Obtained	Groningen (province of Groningen, the Netherlands)
Origin	Geisingen (GFS of Baden-Württemberg, Germany)
Date	September 2004
Collector(s)	R.T.J. Cappers
Context	Shop
GIA-accession	10316
Length/Diameter	–
Thickness	–
Weight	–
Quantity	–
Photographer	D. Fennema

Composition

Triticum aestivum ssp. *spelta*	Spelt	grain kernel

Description
The roasted whole grain kernels have become puffed. As a result, the bran has become fragmented, lending the puffed grain a spotted appearance. The roasted grain is produced by the company Stadtmühle Geisingen under the trade name Dinkel-pops.

Nutritional value (100 g)

carbohydrates	72.4 g
dietary fibre	7.7 g
lipids	2.0 g
minerals	1.8 g
proteins	16.3 g

FIGURE 102

Galiyya

Obtained	Konya (Konya province, Turkey)
Origin	–
Date	May 2015
Collector(s)	R. T. J. Cappers
Context	Shop
GIA-accession	38320
Length/Diameter	–
Thickness	–
Weight	–
Quantity	–
Photographer	D. Fennema

Composition

Cannabis sativa	Hemp	fruit
Triticum aestivum ssp. *aestivum*	Bread wheat	grain kernel

Impurities

Aegilops	Goat beard	rachillate spikelet
Cannabis sativa	Hemp	pedicel
Convolvulus arvensis	Field bindweed	seed
Hordeum vulgare ssp. *vulgare*	Hulled 6-row Barley	floret
Panicum miliaceum	Common millet	floret
Secale cereale	Rye	grain kernel
Triticum aestivum ssp. *aestivum*	Bread wheat	grain kernel (burnt)
Triticum aestivum ssp. *aestivum*	Bread wheat	rachis fragment

Description

The mixture consists of roasted unpolished grain kernels and unroasted fruits of hemp. The fruit wall has small scorch marks that are often also puffed. It also happens that the bran layers have burst open, whereby the starchy endosperm is exposed. The fruits of hemp have been salted separately, and the fruit wall is covered with crusts of salt crystals. Similar salt crystals are also present on the few pedicles of hemp. The impurities consist of a few burnt grain kernels of bread wheat; remains of barley, common millet, and rye; and some diaspores of arable weeds. The grain kernels of rye are dark brown to blackish in colour, and this indicates that they become more easily roasted than those of bread wheat. The mixture is offered for sale in prepackaged units. The Turkish word for this product is '*buğday kavurga*', which translates as 'wheat roasted'.

FIGURE 103

Galiyya

Obtained	Tehran (Tehran province, Iran)
Origin	Sanandaj (Kurdistan province, Iran)
Date	July 2017
Collector(s)	R. T. J. Cappers & M. Kaffashi
Context	Private house
GIA-accession	40793
Length/Diameter	–
Thickness	–
Weight	–
Quantity	–
Photographer	D. Fennema

Composition

Pistacia terebinthus	Turpentine tree	fruit
cf. *Prunus armeniaca*	Apricot	seed
Triticum aestivum ssp. *aestivum*	Bread wheat	grain kernel

Description

The mixture consists of roasted unpolished wheat grain, roasted fruits of the turpentine tree (name in Kurdish: kizwan), and a small admixture of seeds of probably apricot. The seeds of apricot are not roasted and, judging from their bright colour, probably dyed. This can be done naturally with Turmeric (Curcuma longa) or Saffron (Crocus sativus). Scorch marks are present, especially on the grain. The mixture was home-made by the mother of the daughter shown here with a certain quantity of the mixture in a glass bottle (figure on page 132). The daughter did not know how the grain and Turpentine fruits had been roasted. Roasted grain is not often offered for sale in markets.

FIGURE 104

Galiyya

Obtained	Göcer (Karaman province, Turkey)
Origin	–
Date	November 2017
Collector(s)	Y. Okur
Context	Private house
GIA-accession	41066
Length/Diameter	–
Thickness	–
Weight	–
Quantity	–
Photographer	D. Fennema

Composition

Cannabis sativa	Hemp	fruit
Pistacia terebinthus	Turpentine tree	fruit
Triticum aestivum ssp. aestivum	Bread wheat	grain kernel

Impurities

Cicer arietinum	Chick pea	seed
Pistacia terebinthus	Turpentine tree	pedicel

Description

In this sample, the roasted wheat and hemp are only a minor admixture. The presence of a single seed of chick pea is considered unintended. The wheat grain originates from a market in Karaman, whereas the fruits of the turpentine tree are collected from trees growing in the vicinity of Göcer. The roasting of both fruits is done at home. Before roasting, the wheat kernels are soaked in water, and after roasting, they are sprinkled with salt water. Due to the roasting, some of the fruits of the turpentine tree have become inflated and the fruit wall has become smooth.

FIGURE 105

9
Frikeh

Frikeh is an Arabic name for roasted unripe grain kernels. Initially, the harvesting of unripe grain would have been aimed at maximizing the yield. Looking at the domestication history of barley and wheat, we can recognize two stages during which the production of frikeh makes sense in this respect.

The earliest production of frikeh occurred during the stage when only wild cereals were available. It involved the harvesting of unripe grain , which is possible by uprooting the crop or by using harvesting tools, such as sickles. Later production of frikeh was most probably initiated during the stage when naked cereals became available. A shift from hulled to naked cereals forced farmers to deal with increasing yield loss because of the change of what constitutes the threshing unit (from a rachillate triplet [barley] and rachillate spikelet [wheat] to a grain kernel). Maximizing the yield of naked cereals can be realized by, among other things, harvesting at an earlier stage of grain ripening.

The appreciation by consumers of the characteristic taste and smell of frikeh may explain why this traditional food is also produced when yield loss is not problematic (viz. in harvesting hulled domesticated barley and wheat). Today, substantial parts of cereal fields in such countries as Turkey, Syria, Jordan, Lebanon, and Palestine are still harvested at an unripe stage for frikeh production. Modern marketing of frikeh is also focused on its presumed healthy appearance. Egyptian samples of frikeh are even treated with additional colouring to intensify its characteristic green appearance.

Most of the modern frikeh samples held in the collection at GIA concern grain that has been coarsely ground after roasting. Several frikeh samples are characterized by impurities that include diaspores of arable weeds and soil particles. The presence of such impurities is indicative of traditional crop processing in which plants have been cut close to the ground in order to use the straw and the chaff as a source of fuel. When natural gas is used as a fuel, only the infructescences are processed, and hence the proportion of impurities is much reduced.

Frikeh

Obtained	Berlin (GFS of Berlin, Germany)
Origin	–
Date	December 2016
Collector(s)	R. T. J. Cappers
Context	Shop
GIA-accession	39919
Length/Diameter	–
Thickness	–
Weight	–
Quantity	–
Photographer	D. Fennema

Composition

Triticum aestivum ssp. *spelta*	Spelt	grain kernel

Description

The roasted grain kernels have not been ground. The grain was harvested during the milk-ripe stage or the following yellow-ripe stage. Immediately after harvesting, the unripe grain was roasted by using wood of Beech (*Fagus sylvatica*) as fuel. The use of extra fuel, instead of the straw of the whole plant, together with the absence of impurities, such as diaspores of arable weeds, implies that, most probably, only the spikes were roasted. The smoke of the beech contributes to the characteristic aroma of the frikeh. The frikeh is offered for sale as an organically farmed food item, and the German word for this product is '*Grünkern*', which translates as 'green kernel'.

Nutritional value (100 g)

carbohydrates	63.00 g
lipids	2.70 g
proteins	12.00 g
salt	0.07 g

FIGURE 106

Frikeh

Obtained	Amsterdam (province of Noord Holland, the Netherlands)
Origin	Syria
Date	January 2016
Collector(s)	R. T. J. Cappers
Context	Shop
GIA-accession	39746
Length/Diameter	–
Thickness	–
Weight	–
Quantity	–
Photographer	D. Fennema

Composition

Triticum aestivum ssp. *aestivum*	Bread wheat	grain kernel

Impurities		
Avena fatua	Wild oat	floret
Coriandrum sativum	Coriander	seed
Galium	Bedstraw	fruit (mericarp)
Hordeum vulgare ssp. *vulgare*	Hulled 6-row Barley	floret
Lathyrus/Vicia	Pea/Vetch	seed
Lens culinaris	Lentil	seed
Sorghum bicolor	Sorghum	fruit
Sorghum bicolor	Sorghum	spikelet
Silybum marianum	Milk thistle	fruit
Triticum aestivum ssp. *aestivum*	Bread wheat	culm fragment
Triticum aestivum ssp. *aestivum*	Bread wheat	rachis fragment
Triticum turgidum ssp. *durum*	Hard wheat	grain kernel
–	dung	
–	stone	

Description

A coarse-ground frikeh made of unpolished grain kernels. Scorch marks are present. The presence of the rachis fragments, as well as the shape of the grain kernels, is indicative of Bread wheat. But a small admixture of Hard wheat seems to be present. Other impurities consist of vegetative parts of Bread wheat, Barley, Lentil, and Sorghum; diaspores of arable weeds; a few stones; and dung particles. Barley is present in the form of two cultivars, having straw-coloured chaff and grey-coloured chaff. The rather abundant grain kernels of Sorghum, as well as the polished seeds of Lentil and Coriander, do not show any scorch marks, and it seems that they have become mixed with the frikeh after its processing. Its origin is not mentioned on the packaging, but has been confirmed by the importer.

FIGURE 107

Frikeh

Obtained	Amsterdam (province of Noord Holland, the Netherlands)
Origin	Lebanon
Date	January 2016
Collector(s)	R. T. J. Cappers
Context	Shop
GIA-accession	39749
Length/Diameter	–
Thickness	–
Weight	–
Quantity	–
Photographer	D. Fennema

Composition

Triticum aestivum ssp. *aestivum*	Bread wheat	grain kernel
Triticum turgidum ssp. *durum*	Hard wheat	grain kernel

Impurities		
Avena	Oats	floret and fruit
Avena fatua	Wild oat	floret
cf. *Carduus*	Thistles	fruit
Convolvulus	Field bindweed	seed
Coriandrum sativum	Coriander	seed
Fabaceae	Pea family	seed
Galium aparine/tricornutum	(Corn) cleavers	fruit (mericarp)
Hordeum marinum	Sea barley	spikelet and fruit
Hordeum vulgare ssp. *vulgare*	Hulled 6-row Barley	floret
Hordeum/Triticum	Barley/wheat	culm
Lens culinaris	Lentil	seed
Lolium multiflorum	Italian rye-grass	floret
Petroselinum crispum	Parsley	fruit (mericarp)
Phalaris minor	Lesser canary-grass	floret
Phalaris paradoxa	Awned canary-grass	spikelet and floret
Rhus coriaria	Sumac	fruit (endocarp)
Silybum marianum	Milk thistle	fruit
Sorghum halepense	Johnson grass	spikelet
Trigonella foenum-graecum	Fenugreek	seed
Triticum aestivum ssp. *aestivum*	Bread wheat	fruit and rachis fragment
unidentified	fruit and seed	
–	insect	
–	dung	
–	clay and stone	

Description

The frikeh is a mixture of unripe, unpolished grain kernels of Bread wheat and Hard wheat. Both cereals are naked, and it is therefore possible that they were cultivated as a maslin (multiple cropping, usually with a combination of two cereals). Another option is that harvests from different fields were combined. Most of the grain kernels are green. The impurities consist of a variety of economic plants, including Coriander, Barley, Lentil, Parsley, Sumac, and Fenugreek; diaspores of different arable weeds; insects; dung particles; and some clay and stone.

Nutritional value (100 g)

carbohydrates	66 g
lipids	9 g
proteins	14 g

FIGURE 108

Frikeh

Obtained	Amsterdam (province of Noord Holland, the Netherlands)
Origin	Lebanon
Date	February 2016
Collector(s)	Y. Okur
Context	Shop
GIA-accession	39877
Length/Diameter	–
Thickness	–
Weight	–
Quantity	–
Photographer	D. Fennema

Composition

Triticum aestivum ssp. *aestivum*	Bread wheat	grain kernel

Impurities		
Avena	Oat	floret
Avena	Oat	fruit
Avena fatua	Wild oat	floret
Convolvulus arvensis	Field bindweed	seed
Elytrigia repens	Common couch	floret
Galium	Bedstraw	fruit (mericarp)
Hordeum	Wild barley	rachillate triplet
Hordeum vulgare ssp. *distichon*	Hulled 2-row Barley	rachillate triplet
Hordeum vulgare ssp. *vulgare*	Hulled 6-row Barley	floret
Lens culinaris	Lentil	seed
Phalaris	Canary-grass	floret
Phalaris paradoxa	Awned canary-grass	spikelet
Sesamum indicum	Sesame	seed
Silybum marianum	Milk thistle	fruit
Triticum aestivum ssp. *aestivum*	Bread wheat	culm fragment
Triticum aestivum ssp. *aestivum*	Bread wheat	hull
Triticum aestivum ssp. *aestivum*	Bread wheat	rachis fragment
–	insect	
–	stone	

Description

The unpolished grain kernels have been coarsely ground, and some of the grain kernels are still intact. Scorch marks are present, though in a low frequency. Impurities include vegetative plant parts of Bread wheat; other economic plants, including Barley, Lentil, and Sesame; diaspores of arable wheats; insects (beetles); and stones. The florets of Barley are either straw-coloured or grey-brown and thus represent two landraces. Lentil and Sesame must have become mixed with the frikeh after its production. The infestation with beetles must also have occurred after its production.

FIGURE 109

Frikeh

Obtained	Utrecht (province of Utrecht, the Netherlands)
Origin	Lebanon
Date	February 2017
Collector(s)	R. T. J. Cappers
Context	Shop
GIA-accession	40269
Length/Diameter	–
Thickness	–
Weight	–
Quantity	–
Photographer	D. Fennema

Composition

Triticum aestivum ssp. *aestivum*	Bread wheat	grain kernel
Impurities		
Asteraceae	Daisy family	fruit
Avena fatua	Wild oat	floret
Beta vulgaris	Beet	fruit and exocarp
Galium	Bedstraw	fruit (mericarp)
Hordeum vulgare ssp. *vulgare*	Hulled 6-row Barley	floret
Phalaris cf. *paradoxa*	Awned canary-grass	floret
Triticum aestivum ssp. *aestivum*	Bread wheat	grain kernel (fungus)
Triticum aestivum ssp. *aestivum*	Bread wheat	rachis fragment
Vaccaria hispanica	Cowherb	seed
–	insect	
–	stone	

Description

The sample contains both unripe and ripe grain kernels. The unripe grain kernels are greenish in colour and somewhat elongated, whereas the ripe ones have an orange-brown colour and are a bit shorter and broader. Both are characterized by a well-developed tuft of 'hair' on top. The impurities consist of chaff remains still attached to the grain kernels, grain kernels that are infested with a fungus, florets of Barley, and diaspores of some arable weeds. The florets of Barley are represented by two different colours, indicating the presence of two landraces. The shelf life according to the packaging is 2 years. The packaging is of a high quality, adapted to requirements for transport and storage.

FIGURE 110

Frikeh

Obtained	Amsterdam (province of Noord Holland, the Netherlands)
Origin	Jordan
Date	December 2017
Collector(s)	R. T. J. Cappers & Y. Okur
Context	Shop
GIA-accession	41099
Length/Diameter	–
Thickness	–
Weight	–
Quantity	–
Photographer	D. Fennema

Composition

Triticum aestivum ssp. *aestivum*	Bread wheat	grain kernel

Impurities

Avena	Oat	floret
Avena	Oat	fruit
Elytrigia repens	Common couch	floret
Hordeum vulgare ssp. *vulgare*	Hulled 6-row Barley	floret
Malva	Mallow	fruit (mericarp)
Phalaris paradoxa	Awned canary-grass	spikelet
Triticum turgidum ssp. *durum*	Hard wheat	grain kernel
–	stone	

Description

Only some of the unpolished grain kernels have been fragmented during grinding. The impurities concern grain kernels of Hard wheat, florets of Barley, diaspores of some arable weeds, and stones. The shelf life is 2 years according to the packaging.

FIGURE 111

Frikeh

Obtained	Cairo (Cairo governorate, Egypt)
Origin	Beni Suef (Beni Suef governorate, Egypt)
Date	August 2017
Collector(s)	A. Shahat
Context	Shop
GIA-accession	40989
Length/Diameter	–
Thickness	–
Weight	–
Quantity	–
Photographer	D. Fennema

Composition

Triticum aestivum ssp. *aestivum*	Bread wheat	grain kernel

Impurities

Avena fatua	Wild oat	floret
Hordeum vulgare ssp. *vulgare*	Hulled 6-row Barley	floret
–	stone	

Description

A coarse-ground frikeh made of unpolished grain kernels. A proportion of the grain kernels are still intact. Some grain kernels have small scorch marks on their surface. The frikeh has been artificially coloured as well, with an unknown green substance. Small, pulverized units of this green substance are clearly visible in the tuft of 'hair' at the top of the grain kernel. The chaff of the barley and the wild oat are also green-coloured, whereas they have a different life cycle and are usually straw-coloured within a frikeh sample. A similar artificially coloured sample was collected in Khartoum (February 2018). This frikeh originated from Egypt as well.

FIGURE 112

Frikeh

Obtained	Düzce (Düzce province, Turkey)
Origin	Gaziantep (Gaziantep province, Turkey)
Date	February 2017
Collector(s)	R. T. J. Cappers, H. Balcı & L. Kutlu
Context	Farmers market
GIA-accession	40285
Length/Diameter	–
Thickness	–
Weight	–
Quantity	–
Photographer	D. Fennema

Composition

Triticum turgidum ssp. *durum*	Hard wheat	grain kernel
Impurities		
Avena	Oat	floret
Galium	Bedstraw	fruit (mericarp)
Lens culinaris	Lentil	seed
Hordeum vulgare ssp. *vulgare*	Hulled 6-row Barley	floret
Triticum turgidum ssp. *durum*	Hard wheat	chaff
Triticum turgidum ssp. *durum*	Hard wheat	culm fragment
–	insect	
–	stone	

Description

Only some of these unpolished grain kernels have fragmented during grinding. Scorch marks are rare. Impurities consist of some vegetative parts of hard wheat, a seed of Lentil, some florets of Barley, diaspores of arable weeds, stones, and insects. The frikeh has become infested with three different species of grain beetle during storage.

FIGURE 113

Frikeh

Obtained	Istanbul (Istanbul province, Turkey)
Origin	–
Date	February 2017
Collector(s)	R. T. J. Cappers
Context	Shop
GIA-accession	40275
Length/Diameter	–
Thickness	–
Weight	–
Quantity	–
Photographer	D. Fennema

Composition

Triticum turgidum ssp. *durum*	Hard wheat	grain kernel

Impurities

Avena	Oat	floret
Avena	Oat	fruit
Hordeum vulgare ssp. *vulgare*	Hulled 6-row Barley	floret
Lolium temulentum	Darnel	floret
Triticum turgidum ssp. *durum*	Hard wheat	chaff
Triticum turgidum ssp. *durum*	Hard wheat	grain kernel (fungus)
–	stone	

Description

A coarse-ground frikeh made of unpolished grain kernels. A proportion of the grain kernels are still intact. Scorch marks are present, though in a low frequency. The shelf life is 2 years according to the packaging. The Turkish term for this product is '*firik bulgur*', which translates as '*frikeh bulgur*'.

FIGURE 114

Frikeh

Obtained	Istanbul (Istanbul province, Turkey)
Origin	Gaziantep (Gaziantep province, Turkey)
Date	September 2014
Collector(s)	R. T. J. Cappers & Y. Okur
Context	Shop
GIA-accession	35009
Length/Diameter	–
Thickness	–
Weight	–
Quantity	–
Photographer	D. Fennema

Composition

Triticum turgidum ssp. *durum*	Hard wheat	grain kernel

Impurities

Hordeum vulgare ssp. *vulgare*	Hulled 6-row Barley	floret
Triticum turgidum ssp. *durum*	Hard wheat	chaff
Triticum turgidum ssp. *durum*	Hard wheat	grain kernel (fungus)
unidentified	–	culm fragment
Vaccaria hispanica	Cowherb	seed
–	stone	

Description

A coarse-ground frikeh made of unpolished grain kernels. The presence of both large, whole grain kernels and small fragments suggest that the frikeh may contain grains of different origins. Some fragments have clear scorch marks. The colour, which varies from green to more straw-coloured, indicates that part of the grain was ripe during harvesting. The impurities consist of some chaff, a single grain kernel that is undeveloped due to a fungus infection, and diaspores of an arable weed.

FIGURE 115

Frikeh

Obtained	Amsterdam (province of Noord Holland, the Netherlands)
Origin	Gaziantep (Gaziantep province, Turkey)
Date	December 2017
Collector(s)	R. T. J. Cappers & Y. Okur
Context	Shop
GIA-accession	41101
Length/Diameter	–
Thickness	–
Weight	–
Quantity	–
Photographer	D. Fennema

Composition

Triticum turgidum ssp. *durum*	Hard wheat	grain kernel

Impurities		
Avena	Oat	floret
Avena	Oat	fruit
Echium	Viper's buglossus	fruit (mericarp)
Elytrigia repens	Common couch	floret
Fabaceae spp.	Pea family	seed
Galium	Bedstraw	fruit (mericarp)
Hordeum vulgare ssp. *vulgare*	Hulled 6-row Barley	floret
Lens culinaris	Lentil	seed
Lolium temulentum	Darnel	floret
Secale cereale	Rye	grain kernel
Triticum turgidum ssp. *durum*	Hard wheat	culm fragment
Triticum turgidum ssp. *durum*	Hard wheat	chaff
Triticum turgidum ssp. *durum*	Hard wheat	grain kernel (fungus)
–	stone	

Description

A coarse-ground frikeh made of unpolished grain kernels. Not all the grain kernels are green in colour, indicating that the crop was harvested when a proportion of the grain kernels were already ripe at which point they are yellow in colour. A small proportion of the grain kernels have clear scorch marks. A proportion of the grain kernels are unfragmented, and a proportion have been fragmented, either only split lengthwise or also further split crosswise. The shelf life is 2 years according to the packaging.

FIGURE 116

Frikeh

Obtained	Amsterdam (province of Noord Holland, the Netherlands)
Origin	Gaziantep (Gaziantep province, Turkey)
Date	December 2017
Collector(s)	R. T. J. Cappers & Y. Okur
Context	Shop
GIA-accession	41100
Length/Diameter	–
Thickness	–
Weight	–
Quantity	–
Photographer	D. Fennema

Composition

Triticum turgidum ssp. *durum*	Hard wheat	grain kernel

Impurities

Avena	Oat	floret
Avena	Oat	fruit
Fabaceae	Pea family	seed
Galium	Bedstraw	fruit (mericarp)
Hordeum vulgare ssp. *vulgare*	Hulled 6-row Barley	floret
Lithospermum arvense	Field gromwell	fruit (mericarp)
Triticum turgidum ssp. *durum*	Hard wheat	grain kernel (fungus)
–	stone	

Description

The frikeh is made of unpolished grain kernels. The grain kernels have been coarsely ground, and a substantial proportion of this sample consists of intact grain kernels. Most of the grain kernels are greenish, indicating that the crop was harvested at the proper time for frikeh production. Scorch marks are not prominent. The shelf life mentioned on the packaging is 2 years. The impurities concern of a fragment of a wheat grain kernel that is infected with a fungus, some florets of Barley, some diaspores of arable weeds, and stones.

FIGURE 117

Frikeh

Obtained	Groningen (province of Groningen, the Netherlands)
Origin	Gaziantep (Gaziantep province, Turkey)
Date	September 2017
Collector(s)	R. T. J. Cappers
Context	Shop
GIA-accession	40981
Length/Diameter	–
Thickness	–
Weight	–
Quantity	–
Photographer	D. Fennema

Composition

Triticum turgidum ssp. *durum*	Hard wheat	grain kernel

Impurities		
Avena	Oat	floret
Avena	Oat	fruit
Brassicaceae	Cabbage family	pedicel
Fabaceae spp.	Pea family	seed
Galium	Bedstraw	fruit (mericarp)
Hordeum vulgare ssp. *vulgare*	Hulled 6-row Barley	floret
Lolium	Rye-grass	floret
Lolium temulentum	Darnel	floret
Triticum turgidum ssp. *durum*	Hard wheat	culm fragment
Triticum turgidum ssp. *durum*	Hard wheat	rachis fragment
Vaccaria hispanica	Cowherb	seed
–	stone	

Description

A coarse-ground frikeh made of unpolished grain kernels. Few scorch marks are present. Impurities consist of vegetative plant parts of Hard wheat, florets of Barley, diaspores of arable weeds (including quite a number of members of the Pea family), and also quite a number of stones.

FIGURE 118

Frikeh

Obtained	Derik (Al-Hasakah governorate, Syria)
Origin	–
Date	October 2014
Collector(s)	D. Alshawish
Context	Private house
GIA-accession	35088
Length/Diameter	–
Thickness	–
Weight	–
Quantity	–
Photographer	D. Fennema

Composition

Triticum turgidum ssp. *durum*	Hard wheat	grain kernel
Impurities		
Avena fatua	Wild oat	floret
Elytrigia repens	Common couch	floret
Hordeum vulgare ssp. *vulgare*	Hulled 6-row Barley	floret
Phalaris paradoxa	Awned canary-grass	spikelet
Triticum	Wheat	culm fragment
Triticum	Wheat	grain kernel (infected with fungus)
Triticum aestivum ssp. *aestivum*	Bread wheat	grain kernel
Triticum aestivum ssp. *aestivum*	Bread wheat	rachis fragment
Triticum turgidum ssp. *durum*	Hard wheat	hulls
Triticum turgidum ssp. *durum*	Hard wheat	rachis fragment
–	stone	

Description

A coarse-ground frikeh made of unpolished grain kernels. The grain kernels have split lengthwise and in some cases also crosswise. Smaller specimens have not fragmented. Only a minor proportion of the grain kernel has a green colour, indicating that the crop was harvested in a final ripening stage rather than the preceding green stage. Scorch marks are rare. Some grain kernels show damage caused by insects. The impurities consist of plant remains of Bread wheat and Barley and diaspores of some arable weeds. The florets of barley are either straw-coloured or grey-brown and thus represent two landraces. The proportion of roasted ripe grain kernels (galiyya) of Bread wheat is considerable. They have black scorch marks and have mostly broken. It is possible that they represent the crop that was previously growing on that parcel of land and that they were harvested in a ripe stage, together with the still unripe Hard wheat. It is also possible that the frikeh production took place on the same spot where Bread wheat had been roasted earlier.

FIGURE 119

Frikeh

Obtained	Amsterdam (province of Noord Holland, the Netherlands)
Origin	Palestine
Date	January 2016
Collector(s)	R. T. J. Cappers
Context	Shop
GIA-accession	39743
Length/Diameter	–
Thickness	–
Weight	–
Quantity	–
Photographer	D. Fennema

Composition

Triticum turgidum ssp. *durum*	Hard wheat	grain kernel

Impurities		
Galium	Bedstraw	fruit (mericarp)
Hordeum vulgare ssp. *vulgare*	Hulled 6-row Barley	floret
Phalaris paradoxa	Awned canary-grass	spikelet
Poaceae	Grass family	floret
Triticum turgidum ssp. *durum*	Hard wheat	rachis fragment
–	insect	
–	stone	

Description

A medium-ground frikeh made of unpolished grain kernels. The impurities consist of some rachis fragments of hard wheat, florets of barley, some diaspores of arable weeds and a few insects and stones.

FIGURE 120

Frikeh

Obtained	Amsterdam (province of Noord Holland, the Netherlands)
Origin	Jordan
Date	January 2016
Collector(s)	R. T. J. Cappers
Context	Shop
GIA-accession	39744
Length/Diameter	–
Thickness	–
Weight	–
Quantity	–
Photographer	D. Fennema

Composition

Triticum turgidum ssp. *durum*	Hard wheat	grain kernel

Impurities		
Avena	Oat	fruit
Avena fatua	Wild oat	floret
Hordeum vulgare ssp. *vulgare*	Hulled 6-row Barley	
Malva nicaeensis/parviflora	French/Least mallow	fruit (and mericarp)
Triticum turgidum ssp. *durum*	Hard wheat	rachis fragment
–	stone	

Description

A coarse-ground frikeh made of unpolished grain kernels. Scorch marks are frequently present. The impurities consist of some rachis fragments of hard wheat, florets of barley, some diaspores of arable weeds, and a few stones. The frikeh was being sold in quantities of 650 g.

FIGURE 121

Frikeh

Obtained	Amsterdam (province of Noord Holland, the Netherlands)
Origin	Jordan
Date	January 2016
Collector(s)	R. T. J. Cappers
Context	Shop
GIA-accession	39747
Length/Diameter	–
Thickness	–
Weight	–
Quantity	–
Photographer	D. Fennema

Composition

Triticum turgidum ssp. *durum*	Hard wheat	grain kernel

Impurities		
Avena	Oat	floret
cf. *Glaucium*	Horned-poppy	seed
Hordeum vulgare	Hulled Barley	floret
–	insect	
–	stone	

Description

A medium-ground frikeh made of unpolished grain kernels. The frikeh is somewhat moist. Scorch marks are present, though in small proportions. The impurities consist of a single floret of Barley, a diaspore of an arable weed, and some moths. The moths must have gained access to the frikeh during storage.

FIGURE 122

10
Bread

Bread is made by baking unleavened or leavened loaves made from batter or dough. It is the powdery fraction of the grain kernel, produced by pounding or grinding, which is used for mixing with water to produce the batter or dough. When the whole grain kernel is pulverized, meal is obtained. Alternatively, bran and germ can be removed before pulverizing, in which fat-arm four is obtained. The removal of bran and germ can be applied to grain kernels and is called '*polishing*'. In hulled barley, a similar process is called '*pearling*', which combines the removal of the hulls, bran and germ. Polishing and pearling are rather time-consuming and its application is most probably related to the removal of the fats in particular, reduces the risk of mould formation during the final drying process of the baked bread.

The variability of bread is mainly determined by the kind of cereal, the fermentation, and the shaping of the dough. Modern samples of traditionally baked bread show that both barley and wheat are used for bread making. Barley bread is often characterized by the presence of hull fragments, indicating that florets of hulled barley have been ground and that eventually only the coarse fragments of the hulls have been removed by sieving with a meal sieve. Bread made from wheat is characterized by the absence of chaff. The shape of the bread is related to the oven type. If bread is produced for long storage, it has to become dry. This is possible by shaping the dough into flat loaves that become dry by single baking, or by shaping the dough into thick loaves that need to be baked twice. Double baking of bread is possible in a peel oven, in which loaves are put on a horizontal oven floor of a chamber that retains the heat in a very efficient manner. Fermentation was originally achieved by the reuse of small quantities of dough from the previous bread making. Such sourdough bread has a typical slightly acid taste, due to lactic acid fermentation. More recently, special strains of yeast are used for fermentation. If fermentation is applied in flat bread that is produced for long storage, the dough is perforated prior to the baking, a process called '*docking*'. Occasionally, such perforations are still visible in the baked bread.

Bread

Obtained	Amsterdam (province of Noord Holland, the Netherlands)
Origin	–
Date	March 2015
Collector(s)	R. T. J. Cappers & Y. Okur
Context	Private house
GIA-accession	40973
Length/Diameter	(15.2–) 15.9 (–17.2) cm
Thickness	c. 0.5–1.0 cm
Weight	(60.8–) 62.8 (–66.6) g
Quantity	5
Photographer	D. Fennema

Composition

Triticum aestivum ssp. *aestivum*	Bread wheat	grain kernel
–	salt	
–	water	
–	yeast	

Description

A flat bread that was baked on a convex metal plate. This bread was baked in Amsterdam by F. Okur using ingredients and methods similar to those she learned in Turkey (Göcer). The dough is flattened with a rolling pin, on a table. The mobile plate is placed on the kitchen floor and is fired by butane gas. In one session of bread baking, which lasts some 35 minutes, some 40 *mayalı* breads are baked. A mixture of meal and flour is used, which is obtained from a local miller. Today, some of the surplus bread is stored in the refrigerator and some in the freezer. Outside the refrigerator, the *mayalı* breads keep for only a couple of days; in the refrigerator they keep for about a week. Breads that have been stored for a while are improved by briefly reheated them before eating. The Turkish term for this bread is '*mayalı ekmek*', which translates as 'yeast bread'.

FIGURE 123

Bread

Obtained	Nigde (Nigde province, Turkey)
Origin	–
Date	August 2014
Collector(s)	F. Fantone
Context	Private house
GIA-accession	35096
Length/Diameter	18 cm
Thickness	1.4 cm
Weight	158.8 g
Quantity	–
Photographer	D. Fennema

Composition

Triticum	Wheat	grain kernel
–	salt	
–	water	
–	yeast	

Description
A flat bread that was baked on a convex metal plate. The Turkish name is '*bazlama ekmek*', which is a local name within Turkey for this kind of round bread.

FIGURE 124

Bread

Obtained	Chwarqurna (Al-Sulaimaniyah governorate, Iraq)
Origin	–
Date	October 2013
Collector(s)	R. T. J. Cappers
Context	–
GIA-accession	32929
Length/Diameter	34 cm
Thickness	0.1–0.2 cm
Weight	93 g
Quantity	1
Photographer	D. Fennema

Composition

Hordeum vulgare	Barley	grain kernel
–	salt	
–	water	

Description

A flat bread that was baked on a convex metal plate. The dough is flattened with a rolling pin, on a table. This bread was collected for us on request. We were unable to establish whether the bread was made from pure barley meal or if also some wheat meal or flour was added to improve its baking quality. And we were unable to establish whether the bread was baked in a private house or in a bakery. The bread is quite dark in colour and has a rather compact structure, which does not allow for the formation of air bubbles. The round bread was folded three times to obtain a smaller, rectangular unit to facilitate transport.

FIGURE 125

Bread

Obtained	Dara Gulan (Sulaimaniyah governorate, Iraq)
Origin	–
Date	August 2013
Collector(s)	R.T.J.Cappers
Context	Private house
GIA-accession	32938
Length/Diameter	46 cm
Thickness	c. 0.1 cm
Weight	67 g
Quantity	1
Photographer	D.Fennema

Composition

Triticum	Wheat	grain kernel
–	salt	
–	water	
–	yeast	

Description

A flat bread that was baked on a convex metal plate. The dough is flattened with a rolling pin, on a table. The pattern of irregular circles shows where the dough has risen due to the expansion of water vapour and carbon dioxide. These parts of the dough made less contact with the hot plate and are consequently lighter in colour. The round bread was folded three times to obtain a smaller, rectangular unit to facilitate transport. The surface has become somewhat mouldy.

FIGURE 126

Bread

Obtained	Pirdmamka (Al-Sulaimaniyah governorate, Iraq)
Origin	–
Date	October 2013
Collector(s)	R.T.J. Cappers
Context	Private house
GIA-accession	33123
Length/Diameter	c. 100 cm
Thickness	c. 0.2 cm
Weight	117 g
Quantity	1
Photographer	D. Fennema

Composition

Triticum	Wheat	grain kernel
–	salt	
–	water	

Description

A flat bread that was baked on a convex metal plate. The dough is flattened with a rolling pin, on a table. The edge of the bread was heated a bit more than the inner part. The round bread was folded twice to obtain a smaller, square-ish unit to facilitate transport.

FIGURE 127

Bread

Obtained	Chwarqurna (Al-Sulaimaniyah governorate, Iraq)
Origin	–
Date	October 2013
Collector(s)	R. T. J. Cappers
Context	Bakery
GIA-accession	33122
Length/Diameter	c. 36 cm
Thickness	c. 0.2 cm
Weight	72 g
Quantity	1
Photographer	D. Fennema

Composition

Triticum	Wheat	grain kernel
–	salt	
–	water	

Description

A flat bread that was baked on a convex metal plate. The dough ball is shaped into a thin, round unit with a press, which gives the bread its typical waffle pattern. The round bread was folded three times to obtain a smaller, rectangular unit to facilitate transport.

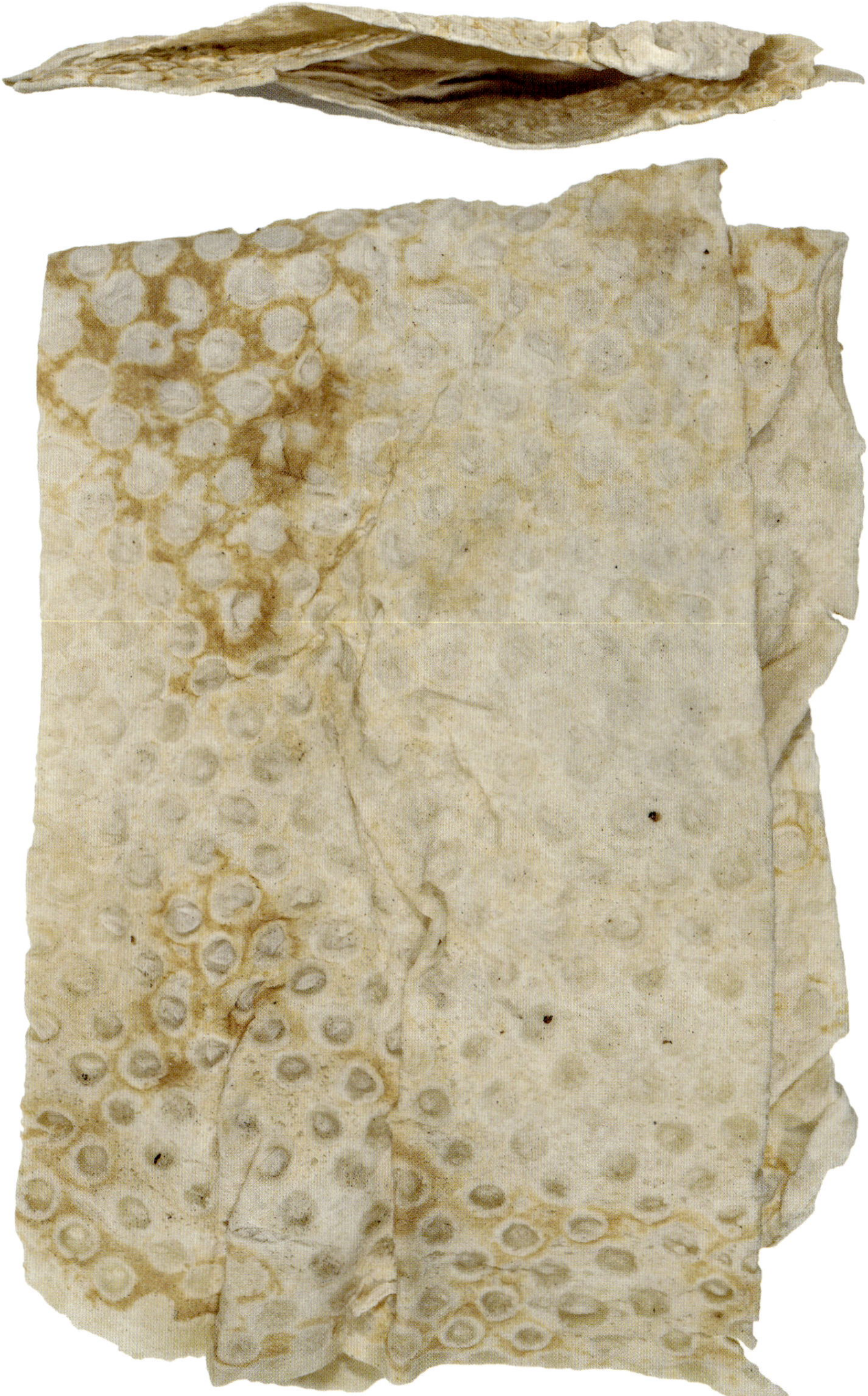

FIGURE 128

Bread

Obtained	Jamma'en (West Bank, Palestine)
Origin	–
Date	September 2015
Collector(s)	H. Madina
Context	Private house
GIA-accession	41112
Length/Diameter	c. 50 cm
Thickness	–
Weight	275 g
Quantity	–
Photographer	D. Fennema

Composition

Triticum	Wheat	grain kernel
–	salt	
–	water	
–	yeast	

Description

A flat bread that was baked in a tabun oven. The dough is flattened using both hands. The flattened dough is spread over the large-sized pebbles on the inner surface of the tabun. The bread is baked for 3–4 minutes. Some 10 breads are baked in a single session, and occasionally the total number can be as much as 50. Such bread may show impressions of large pebbles, especially on the bottom face. In folded bread, such imprints may no longer be visible. The use of large pebbles has the advantage that any pebbles that stick to the baked bread are easily recognized and removed. It was baked by F. Madina. The round bread was folded twice to obtain a smaller, square-ish unit to facilitate transport. Because the tabun oven is heated 24 hours a day, 7 days a week, bread is baked shortly prior to consumption and not kept for a long period of time.

FIGURE 129

Bread

Obtained	Karaman (Karaman province, Turkey)
Origin	–
Date	July 2013
Collector(s)	Y. Okur
Context	Private house
GIA-accession	32928
Length/Diameter	20–22 cm
Thickness	3.0–3.5 cm
Weight	304 g
Quantity	1
Photographer	D. Fennema

Composition

Sesamum indicum	Sesame	seed
Triticum	Wheat	grain kernel
–	egg	
–	salt	
–	water	
–	yeast	

Description

A thick bread that was baked in an above-ground tannur oven. It takes 2 hours for the dough to rise. The upper side of the bread is covered with a mixture of beaten egg white and sesame. The family originates from Diyarbakır (Diyarbakır province) and recreated the kind of oven they were accustomed to operating in their place of origin. The kind of fuel used for baking this bread differs with the seasons. In summer, prunings of apple trees (*Malus sylvestris*) are used, whereas in winter, dung cakes are used. The pre-heating of the oven takes 2 hours. A quantity of 30–40 breads is baked once every 2 weeks.

FIGURE 130

Bread

Obtained	Cariklar Fatih (Konya province, Turkey)
Origin	–
Date	May 2015
Collector(s)	R. T. J. Cappers & Y. Okur
Context	Private house
GIA-accession	41111
Length/Diameter	28 cm
Thickness	2 cm
Weight	408 g
Quantity	1
Photographer	D. Fennema

Composition

Triticum	Wheat	grain kernel
–	salt	
–	water	

Description

A flat bread that was baked in an above-ground tannur oven that had been constructed 20 days earlier by people who were working in this area temporarily as charcoal burners. The bread is sun-dried and can be stored for about 3 weeks. Nevertheless, fresh bread is baked every 3 days.

FIGURE 131

Bread

Obtained	Dara Gulan (Sulaimaniyah governorate, Iraq)
Origin	–
Date	August 2013
Collector(s)	R. T. J. Cappers
Context	Private house
GIA-accession	32930
Length/Diameter	21–24 cm
Thickness	1 cm
Weight	110–111 g
Quantity	3
Photographer	D. Fennema

Composition

Triticum	Wheat	grain kernel
–	salt	
–	water	
–	yeast	

Description

A flat bread that was baked in an above-ground tannur oven. The dough is shaped into slightly elongated loaves. The rising has resulted in an irregular puffing. Scorch marks are present on the upper side of the bread, which is the side that faces the inner part of the oven during baking.

FIGURE 132

Bread

Obtained	Skoura (Daraâ-Tafilalt region, Morocco)
Origin	–
Date	March 2016
Collector(s)	R. T. J. Cappers
Context	–
GIA-accession	41151
Length/Diameter	45 cm
Thickness	1 cm
Weight	460 g
Quantity	1
Photographer	D. Fennema

Composition

Triticum	Wheat	grain kernel
–	salt	
–	water	
–	yeast	

Description

A flat bread that was baked in an above-ground tannur oven. The tannur is located in a small bakehouse that is connected to the other buildings. During the baking process, much smoke becomes concentrated in this house, especially in the upper part. The smoke is less of a nuisance to someone in a crouched position than to someone standing upright. The mixing of the dough is done in the bakehouse. The dough is pushed to the hot, inner part of the tannur, which consists of earthenware. The dough is then flattened by using both hands, resulting in a final shape that is oval and somewhat curved. One or two such breads can be baked at one time.

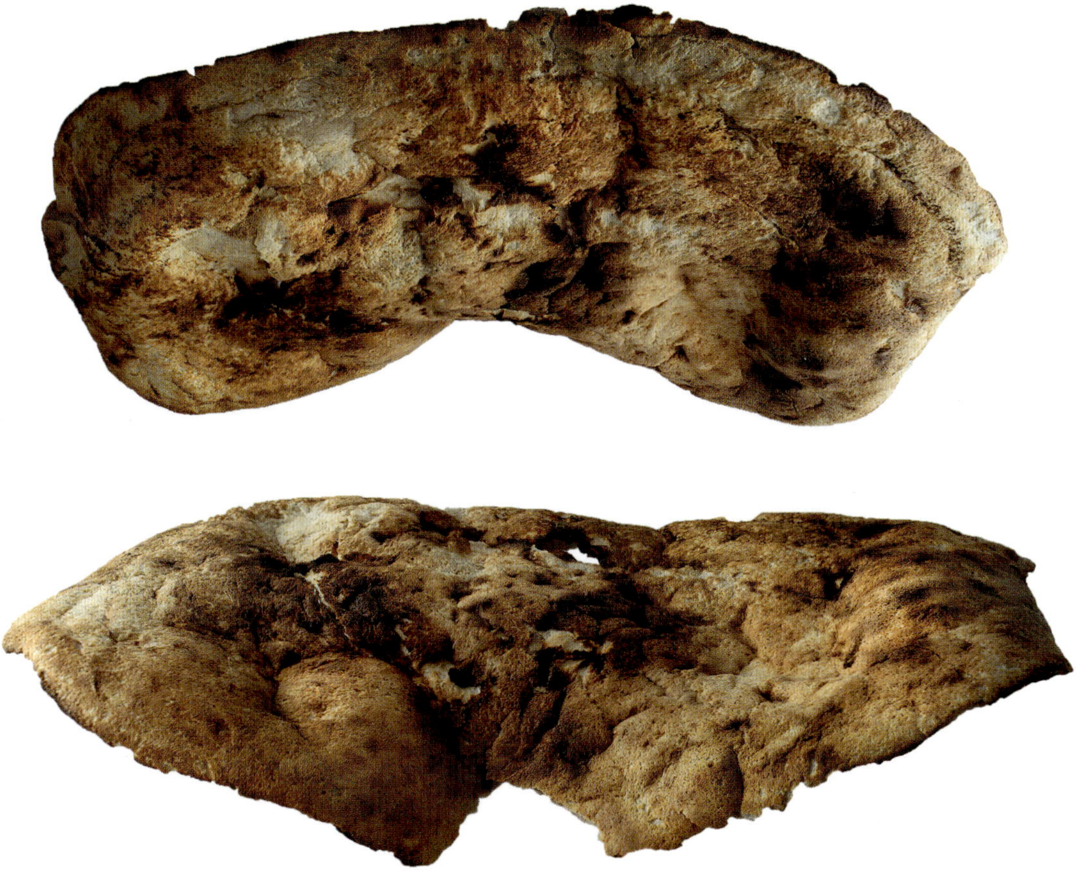

FIGURE 133

Bread

Obtained	Abu Bakra (Al-Sulaimaniyah governorate, Iraq)
Origin	–
Date	June 2014
Collector(s)	R. T. J. Cappers
Context	Private house
GIA-accession	40970
Length/Diameter	11–12 cm
Thickness	2–3 cm
Weight	60 g
Quantity	1
Photographer	D. Fennema

Composition

Triticum	Wheat	grain kernel
–	water	
–	yeast	

Description

A thick bread that is baked in an above-ground tannur oven. This small-sized bread is specially baked for the Anatolian shepherd dog who protects the courtyard. Four of such breads are given to the dog every day: two in the morning and two in the evening. It is baked after finishing baking the bread for human consumption. When this bread making was documented, new fuel had to be added into the oven to increase the temperature again. The dough is made from unsieved wheat flour, water from the stream, and some yeast. The loaves are flattened with the hands on the hot metal plate. There is no effort to create a nice standard shape of bread. The bread is kept in a large plastic bag in a barn.

FIGURE 134

Bread

Obtained	Ilam (Ilam province, Iran)
Origin	–
Date	April 2017
Collector(s)	R. T. J. Cappers & M. Kaffashi
Context	Bakery
GIA-accession	40784
Length/Diameter	19 cm
Thickness	c. 0.5 cm
Weight	91 g
Quantity	3
Photographer	D. Fennema

Composition

Hordeum vulgare	Barley	grain kernel
–	salt	
–	water	
–	yeast	

Description

A flat bread that was single-baked in a direct-heated, single-chambered peel oven. This barley bread is baked in the morning, whereas wheat bread is baked in the afternoon. The dough is mechanically divided and shaped into a flat unit. Next it is manually perforated with a plastic dock that has the shape of a comb. Each punch results in a line of small perforations that may also be seen on the bottom face. This kind of dock is also used by other bakers, but the implement could not be obtained in shops that are specialized in bakery equipment. The Persian name for this bread is '*taftoon*'. The owner of the factory has studied food technology, and she promotes the consumption of bread made with barley.

FIGURE 135

Bread

Obtained	Cairo (Cairo governorate, Egypt)
Origin	–
Date	January 2013
Collector(s)	R.T.J. Cappers
Context	Bakery
GIA-accession	32931
Length/Diameter	15 cm
Thickness	3 cm
Weight	48 g
Quantity	1
Photographer	D. Fennema

Composition

Triticum	Wheat	grain kernel
–	salt	
–	water	

Description
A flat bread that was single-baked in a direct-heated, singled-chambered peel oven. Its surface is covered with bran to prevent the bread from sticking to the substrate. The dough is not docked, and the bread is much inflated when freshly baked. During the cooling process, much of the air escapes and the bread becomes rather flat again. Such breads only have a long shelf life if they are thoroughly dried. The Arabic name for the bread is '*pita*'. It is this kind of bread that is currently the staple food of most Egyptians. Its production is subsidized by the Egyptian government to ensure that each Egyptian citizen can buy a minimum number of these breads every day. But economic recession put pressure on the authorities to reduce this support. A reduction in the number of breads is not an option, as this would result in alarm among consumers, so producers reduce the size of the individual loaves or replace part of the wheat flour with flour with that of Maize (*Zea mays*), which is cheaper.

FIGURE 136

Bread

Obtained	Çaybaşı (Ordu province, Turkey)
Origin	–
Date	August 2015
Collector(s)	R. T. J. Cappers & Y. Okur
Context	Private house
GIA-accession	41114
Length/Diameter	22–24 cm
Thickness	3–5 cm
Weight	680 g
Quantity	1
Photographer	D. Fennema

Composition

Triticum	Wheat	grain kernel
–	salt	
–	water	
–	yeast	

Description
A thick bread that was single-baked in a pre-heated, singled-chambered peel oven. The bread has a somewhat irregular shape. The bottom part is rather black in colour, most probably because the ashes were not completely removed from the oven floor after heating. In this type of oven, the ash is removed to the outside, through the front opening. This has to be done quickly in order to keep as much heat inside as possible. The bottom face of the bread also shows an imprint of the screen on which the bread was dried after baking.

FIGURE 137

Bread

Obtained	Ağsaklı (Konya province, Turkey)
Origin	–
Date	May 2015
Collector(s)	R. T. J. Cappers & Y. Okur
Context	Private house
GIA-accession	41113
Length/Diameter	18–22 cm
Thickness	2.5–4.0 cm
Weight	315 g
Quantity	1
Photographer	D. Fennema

Composition

Triticum	Wheat	grain kernel
–	salt	
–	water	
–	yeast	

Description

A thick bread that was single-baked in a pre-heated, singled-chambered peel oven. This bread is one of the 104 breads that were baked in a single session. The oven provides only enough space for eight such breads, which implies that this number of breads was baked in 13 sessions. The edge of the bread is more leavened than the central part, which may even show a small hole. After baking, the breads are allowed to cool down on a piece of fabric in an adjoining room. The breads will keep for 3 days at room temperature. Nowadays, they are stored in a freezer, where they will keep for about 1 month. The Turkish word for this bread is '*köy ekmeği*', meaning 'village bread'.

FIGURE 138

Bread

Obtained	El Borgeige (River Nile state, Sudan)
Origin	–
Date	January 2014
Collector(s)	R. T. J. Cappers
Context	Bakery
GIA-accession	34572
Length/Diameter	12–13 cm
Thickness	1.5–2.0 cm
Weight	(37.5–) 42.8 (–47.6) g
Quantity	3
Photographer	D. Fennema

Composition

Triticum	Wheat	grain kernel
–	salt	
–	water	
–	yeast	

Description

A flat bread that was single-baked in a direct-heated, singled-chambered peel oven. The oven is heated by gas. Proving takes place after shaping. To prevent the loaves from sticking to the wooden racks on which proving takes place or to the oven floor, the bottoms of the loaves are covered in a coarse fraction of bran. A fine fraction of bran is sprinkled on the tops of the loaves. These kinds of small breads are baked in large quantities in a bakery and can be bought during fixed hours each day.

FIGURE 139

Bread

Obtained	Kalamafka (Crete region, Greece)
Origin	–
Date	July 2016
Collector(s)	R. T. J. Cappers
Context	Private house
GIA-accession	39898
Length/Diameter	23.7 cm
Thickness	8 cm
Weight	650 g
Quantity	1
Photographer	D. Fennema

Composition

Triticum aestivum ssp. *aestivum*	Bread wheat	grain kernel
–	salt	
–	water	
–	yeast	

Description

A thick bread that was single-baked in a pre-heated, double-chambered peel oven. Although the bread is baked in large quantities, it is not baked in the manner of a rusk bread but is intended for prompt consumption. The oven is fuelled with wood. This bread is one of some hundred breads that were baked for a wedding. A small piece of charcoal has stuck to the bottom.

FIGURE 140

Bread

Obtained	Fayum (Fayum governorate, Egypt)
Origin	–
Date	November 2012
Collector(s)	R.T.J. Cappers
Context	Private house
GIA-accession	32933
Length/Diameter	43 cm
Thickness	c. 1 cm
Weight	440–460 g
Quantity	–
Photographer	D. Fennema

Composition

Triticum	Wheat	grain kernel
–	salt	
–	water	
–	yeast	

Description

A flat bread that was twice-baked in a direct-heated, double-chambered peel oven. The dough is allowed to rise overnight and is then divided into smaller units that are covered again to allow for the final rising (proving). The balls are shaped into large, flat units by using a peel with a short handle and a large, circular head. Such peels are traditionally made of the midribs of the leaves of the Date palm (*Phoenix dactylifera*), and the Arabic term for such a peel is '*matraha*'. First, the ball is flattened against the head of the peel using both hands. Next, the loaf is further flattened by tossing the peel frequently, which requires a lot practice. Before the flattened loaf is pushed into the oven with the peel, the loaf is docked to allow carbon dioxide gas to escape, so that the bread remains flat. The home-made peel oven that is used in Egypt is rather small, and only one bread can be baked at a time. The first baking of each bread takes slightly more than 1 minute. Once all of the breads have been baked, the opening of the fuel space is closed to retain the heat in the oven. The breads are cut in quarters to obtain a fresh cut surface that facilitates further loss of water during the second baking process. The twice-baked (rusk) bread can be kept for a long time. The oven was heated with culms of Maize (*Zea mays*). In this way, the principal product of one cereal is heated with the by-product of another cereal. The photo shows two quarters of one of these twice-baked breads.

FIGURE 141

Bread

Obtained	Campobasso (province of Campobasso, Italy)
Origin	–
Date	August 2014
Collector(s)	F. Fantone
Context	Bakery
GIA-accession	41115
Length/Diameter	26 cm
Thickness	c. 0.2 cm
Weight	c. 30 g
Quantity	–
Photographer	D. Fennema

Composition

Triticum turgidum ssp. *durum*	Hard wheat	grain kernel
–	salt	
–	water	
–	yeast (sourdough)	

Description

A flat bread that was twice-baked in a peel oven. The dough is flattened into a paper-thin loaf that is not docked before the first baking. During the first baking, a crust forms and the bread becomes a bit inflated. This makes it possible to cut into the edge after baking and to separate the two sides. These are put back in the peel oven so that the remaining moisture can escape from the bread to make it extremely dry and thus suitable for long-term storage. This kind of twice-baked (rusk) bread used to be produced for shepherds in Sardinia, who took the bread with them during seasonal movements. The bread could be transported vertically in saddlebags. The Italian name is '*pane carasau*', which means 'roasted bread'.

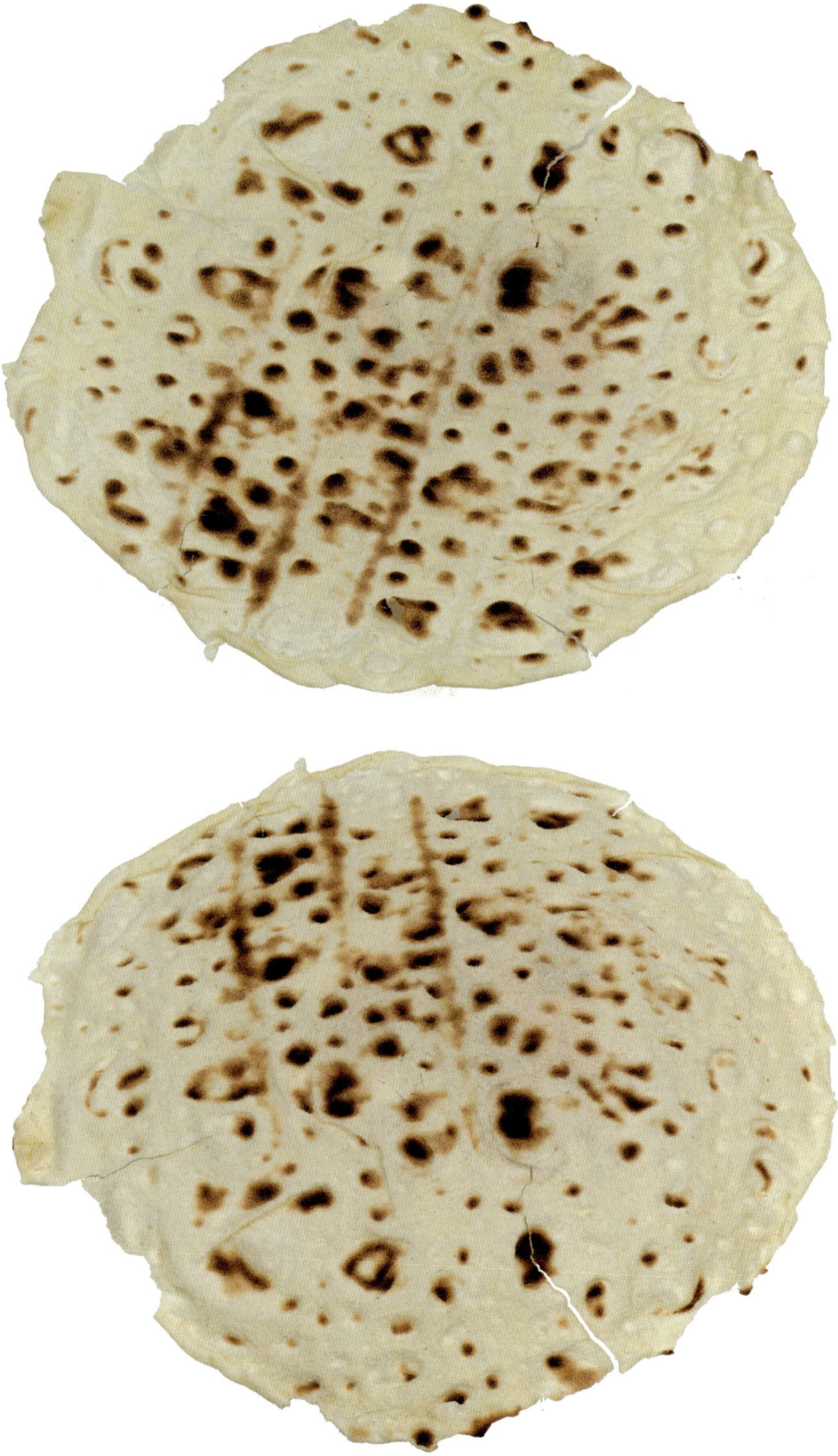

FIGURE 142

Bread

Obtained	Crete (Crete region, Greece)
Origin	–
Date	October 2013
Collector(s)	T. Kalantzo
Context	Private house
GIA-accession	32934
Length/Diameter	10 cm
Thickness	3 cm
Weight	60 g
Quantity	1
Photographer	D. Fennema

Composition

Hordeum vulgare ssp. *distichon*	Hulled 2-row Barley	floret
–	salt	
–	water	
–	yeast	

Description

A thick bread that was twice-baked in a pre-heated, double-chambered peel oven. After the first rising period, the dough is divided into large units that are shaped into elongated loaves. Incisions are made with a knife, and a proving period follows. Once the bread has been baked for the first time, it is allowed to cool down outside the oven. Once the breads are cool enough to handle, the breads are quickly sliced by making use of the previously created incisions. The slices from all the breads are put back into the oven, upon which the openings of both chambers are tightly closed to retain the heat. Moisture can now escape from the fresh cuts, and the slices become extremely dry and thus suitable for long-term storage. The slices cannot be stacked on top of each other, and placing the huge number of slices back into the oven is quite a delicate enterprise. The bread contains a high concentration of hulls (palea and lemma), which indicates that florets rather than grain kernels have been used and that the meal has not been sieved to a high level of purity.

FIGURE 143

Bread

Obtained	Perivolakia (Crete region, Greece)
Origin	–
Date	May 2014
Collector(s)	R.T.J. Cappers
Context	Private house
GIA-accession	34740
Length/Diameter	(10.5–) 12.1 (–13.5) cm
Thickness	(2.5–) 3.9 (–4.8) cm
Weight	(41.7–) 61.0 (–77.5) g
Quantity	6
Photographer	D. Fennema

Composition

Hordeum vulgare	Hulled Barley	floret
–	salt	
–	water	
–	yeast	

Description

A thick bread that was twice-baked in a pre-heated, double-chambered peel oven. The mixing, shaping, and baking of this bread is similar to that of GIA-accession 32934 (page 244). The wheat was imported from France. Before the flour is mixed with the other ingredients, it is sieved with a flour sieve to extract the large fragments of hulls that are still present in the meal. Each bread is sliced into 6–7 slices. The breaks on the bottom crust are somewhat irregular.

FIGURE 144

Bread

Obtained	Kalamafka (Crete region, Greece)
Origin	–
Date	July 2016
Collector(s)	R. T. J. Cappers
Context	Private house
GIA-accession	39905
Length/Diameter	13 cm
Thickness	6.5 cm
Weight	131.9 g
Quantity	1
Photographer	D. Fennema

Composition

Nigella sativa	Black cumin	seed
Triticum aestivum ssp. *aestivum*	Bread wheat	grain kernel
–	salt	
–	water	
–	yeast	

Description

A thick bread that was twice-baked in a pre-heated, double-chambered peel oven. This thick slice of bread was broken from an elongated bread prior to the second baking. Bran fragments are visible on the inside of the bread. The upper part has been sprinkled with seeds of black cumin.

FIGURE 145

Bread

Obtained	Katharon (Crete region; Greece)
Origin	–
Date	July 2016
Collector(s)	R. T. J. Cappers
Context	Private house
GIA-accession	39903
Length/Diameter	8.5 cm
Thickness	3.5 cm
Weight	168 g
Quantity	1
Photographer	D. Fennema

Composition

Hordeum vulgare	Hulled Barley	floret
–	salt	
–	water	
–	yeast	

Description

A thick bread that was twice-baked in a pre-heated, double-chambered peel oven. Incisions are made in the elongated loaves to facilitate the breaking into slices after the first baking. Many small chaff fragments are present; it appears that only the larger ones have been sieved out prior to the mixing of the ingredients.

FIGURE 146

Bread

Obtained	Kritsa (Crete region, Greece)
Origin	–
Date	July 2016
Collector(s)	R. T. J. Cappers
Context	Private house
GIA-accession	39899
Length/Diameter	13.5–14.5 cm
Thickness	7.5 cm
Weight	151–175 g
Quantity	2
Photographer	D. Fennema

Composition

Hordeum vulgare	Hulled Barley	floret
–	salt	
–	water	
–	yeast	

Description

A thick bread that was twice-baked in a pre-heated, double-chambered peel oven. Incisions are made in the elongated loaves to facilitate the breaking into slices after the first baking. These incisions are recognized from the whitish colour of part of the original outside of the now-fractured pieces. The fracture surface, which is only directly exposed to the heat of the oven during the second baking stage, is much darker. Many chaff fragments are present, and it seems as if no sieving took place after grinding. Some small charcoal fragments are present on the bottom of the bread. Such small particles escape being cleaned from the oven floor before the bread is put in the oven with a peel. This bread was given to us by a woman who had finished baking, having made three large bags full of such sliced rusk bread.

FIGURE 147

Bread

Obtained	Thessaloniki (Central Macedonia region, Greece)
Origin	–
Date	June 2013
Collector(s)	R. T. J. Cappers
Context	Private house
GIA-accession	32932
Length/Diameter	6.0–6.5 cm
Thickness	2.8 cm
Weight	14.5–15.4 g
Quantity	2
Photographer	D. Fennema

Composition

Hordeum vulgare	Hulled Barley	floret
Prunus dulcis	Almond	seed
Sesamum indicum	Sesame	seed
Vitis vinifera	Grape	fruit
–	salt	
–	water	
–	yeast	

Description

A thick bread that was twice-baked in a peel oven. Incisions are made in the elongated loaves to facilitate the breaking into slices after the first baking. The barley rusk bread is rather small in size and contains fragments of the chaff. The bread is spiced with polished seeds of sesame, whole fruits of grape, and coarse-ground seeds of almond. The presence of these more fatty ingredients does not affect the quality of the bread: after some four years of storage in a plastic bag there was no sign of it going mouldy. The rusk bread was specially made for the dinner of the International Work Group for Palaeoethnobotany that was organized by T. Valamoti in Thessaloniki, in the summer of 2003.

FIGURE 148

Bread

Obtained	Perivolakia (Crete region, Greece)
Origin	–
Date	May 2014
Collector(s)	R. T. J. Cappers
Context	Private house
GIA-accession	34739
Length/Diameter	18cm
Thickness	2.5 cm
Weight	171 g
Quantity	1
Photographer	D. Fennema

Composition

Hordeum vulgare	Hulled Barley	floret
–	salt	
–	water	
–	yeast	

Description

A thick bread that was twice-baked in a pre-heated, double-chambered peel oven. Many fragments of chaff are present in the bread. Probably only the large fragments were sieved out prior to the mixing of the ingredients. The dough is first rolled into a cylinder with a rolling pin, on a table. Then the pin is pushed into the cylinder to create a groove. Next, the cylinder is bent into a circle and somewhat flattened. After the first baking, the bread is broken crosswise, making use of the presence of the groove. A large knife is used to cut the bread into halves, which are returned to the oven for the second baking.

FIGURE 149

Bread

Obtained	Campobasso (Province of Campobasso, Italy)
Origin	–
Date	September 2015
Collector(s)	F. Fantone
Context	Bakery
GIA-accession	39676
Length/Diameter	16 cm
Thickness	1.5–2.0 cm
Weight	105–110 g
Quantity	3
Photographer	D. Fennema

Composition

Triticum	Wheat	bran
Triticum turgidum ssp. *durum*	Hard wheat	grain kernel
–	salt	
–	water	
–	yeast (sourdough)	

Description

A thick bread that was twice-baked in a peel oven. The Italian label mentions '*farina tipo "0"*', which refers to a soft flour, which needs a moderate rising time, and to '*con crusca*', which means that bran has been added. The hole in the centre may originally have been used for transport and storage, in which breads were strung on a rope or stacked. The local Italian name for this rusk bread is '*freselle*'.

FIGURE 150

Bread

Obtained	Campobasso (province of Campobasso, Italy)
Origin	–
Date	February 2014
Collector(s)	F. Fantone
Context	Bakery
GIA-accession	41116
Length/Diameter	12–15 cm
Thickness	2–3 cm
Weight	70–85 g
Quantity	4
Photographer	D. Fennema

Composition

Triticum turgidum ssp. *durum*	Hard wheat	grain kernel
–	salt	
–	water	
–	yeast (sourdough)	

Description

A thick bread that was twice-baked in a peel oven. Its production is similar to that of GIA-accession 39676 (page 258), but no hole is made in the middle.

FIGURE 151

Bread

Obtained	Aarhus (Central Jutland, Denmark)
Origin	–
Date	June 2015
Collector(s)	K. K. Nielsen
Context	Factory
GIA-accession	38803
Length/Diameter	(6.2–) 6.5 (–6.9) cm
Thickness	(1.6–) 1.9 (–2.3) cm
Weight	(7.4–) 8.9 (–11.6) g
Quantity	8
Photographer	D. Fennema

Composition

Triticum	Wheat	grain kernel
–	ascorbate E 300	
–	food additive E 472e	
–	oil (vegetable)	
–	salt	
–	sugar	
–	water	
–	yeast	

Description

A thick bread that was twice-baked in a peel oven. Food additives consist of Vitamin C (E 300) and an emulsifier consisting of fatty acids (E 472e). The bread is baked into small units that are split before the second baking. The Danish name is *'tvebakker'*, meaning rusks. It is usually eaten with some butter on top. The shelf life according to the packaging is 3 years.

Nutritional value (100 g)

carbohydrates	71 g
dietary fibre	3.3 g
lipids	11 g
proteins	10 g
salt	1.1 g

FIGURE 152

Bread

Obtained	Aarhus (Central Jutland, Denmark)
Origin	–
Date	June 2015
Collector(s)	K. K. Nielsen
Context	Bakery
GIA-accession	38804
Length/Diameter	3.0–3.1 cm
Thickness	1.1 cm
Weight	(3.9–) 4.0 (–4.1) g
Quantity	8
Photographer	D. Fennema

Composition

Citrus medica	Citron	fruit
Daucus carota	Carrot	root
Elettaria cardamomum	Cardamom	seed
Solanum tuberosum	Potato	tuber
Triticum	Wheat	grain kernel
Vanilla planifolia	Vanilla	fruit
–	egg	
–	food additive E 322	
–	oil (vegetable)	
–	salt	
–	sugar	
–	water	
–	yeast	

Description

A thick bread that was twice-baked in a peel oven. This sweet rusk bread is characterized by the addition of quite a number of other plant ingredients, including ones from the New World (potato starch) and Asia (Citron, Cardamon, and Vanilla). This rusk bread is traditionally eaten by putting some units in a buttermilk-based product called '*koldskal*'. The Danish name for this traditional rusk bread is '*kammerjunkere*'.

Nutritional value (100 g)

carbohydrates	70 g
lipids	19 g
proteins	6 g
salt	0.27 g

FIGURE 153

Bread

Obtained	Haren (province of Groningen, the Netherlands) and Medemblik (province of Noord Holland, the Netherlands)
Origin	–
Date	June 2015
Collector(s)	D. Fennema and Y. Okur
Context	Shop and bakery
GIA-accession	41117
Length/Diameter	(8.8–) 9.2 (–9.8) cm
Thickness	(1.2–) 1.6 (–2.0) cm
Weight	(9.4–) 13.5 (–18.0) g
Quantity	11
Photographer	D. Fennema

Composition

Triticum aestivum ssp. *aestivum*	Bread wheat	grain kernel
–	salt	
–	water	
–	yeast	

Description

A thick bread that was baked in a peel oven. This kind of bread is sometimes offered for sale as single-baked bread, but usually it is sliced in two and twice-baked. The upper and middle rows shows a fresh, single-baked bread that was obtained from a shop in Haren. Both sides are rounded and have a soft, brownish crust. The right-hand specimen in the middle row has been cut and shows the soft, white inner part. The specimens in the bottom row have been twice-baked and were obtained from a bakery in Medemblik. The inside of the sliced rusk bread is rough (left) and the outside rather smooth (right). The rusk breads are rather fragile and are kept in special rusk tins, which are cylindrical. The Dutch name for this rusk bread is '*beschuit*'.

FIGURE 154

11
Bulgur

Bulgur is dried par-boiled and granulated grain. The polishing of the grain before granulation is optional. Most probably, par-boiling has been applied in first instance to hulled wheat to facilitate the removal of the chaff. In this line of reasoning, bulgur can be considered as a traditional food originally made from hulled wheat taxa such as Einkorn (*Triticum monococcum* ssp. *monococcum*), Emmer (*Triticum turgidum* ssp. *dicoccon*) and Spelt (*Triticum aestivum* ssp. *spelta*).

Further fragmentation of the par-boiled grain by either pounding or grinding produces a mixture of different size fractions. To make use of a consistent cooking time, such mixtures are separated into different grades and each grade is used for making specific dishes. Bulgur also has become an ingredient of tarhana, which is a mixture of grain and milk.

Because hulled wheat has been replaced by naked wheat in recent times on a large scale, bulgur is nowadays made from naked wheat, such as Hard wheat (*Triticum turgidum* ssp. *durum*) and Bread wheat (*Triticum aestivum* ssp. *aestivum*). This illustrates that bulgur is much appreciated as a traditional food and that par-boiling is still applied whereas its original function has been lost.

The modern samples of bulgur are made from either hulled wheat or naked wheat. The ones from hulled wheat represent the original bulgur and have been collected in Turkey and Iran. Further variation of the bulgur samples deals with the particle size of the bulgur and the polishing that can be applied prior to fragmentation.

Bulgur

Obtained	Çaybaşi (Ordu province, Turkey)
Origin	–
Date	August 2015
Collector(s)	R.T.J. Cappers & Y. Okur
Context	Farmhouse
GIA-accession	39632
Length/Diameter	–
Thickness	–
Weight	–
Quantity	–
Photographer	D. Fennema

Composition

Triticum monococcum ssp. *monococcum*	Einkorn	grain kernel

Impurities

Agrostemma githago	Corncockle	seed
Apiaceae	Carrot family	fruit (mericarp)
Avena	Oat	fruit
Avena fatua	Wild oat	floret
Elytrigia repens	Common couch	floret
Fabaceae	Pea family	seed
Lolium temulentum	Darnel	floret
Melampyrum arvense	Field cow-wheat	seed
Ranunculus arvensis	Corn buttercup	fruit
Secale cereale	Rye	grain kernel
Triticum monococcum ssp. *monococcum*	Einkorn	rachillate spikelet
–	stone	

Description

A coarse-ground bulgur made of unpolished grain kernels. The bulgur is home-made. The rachillate spikelets have been par-boiled prior to the removal of the rachis segments and the chaff (de-graining). Small fragments of the lemma and palea are still attached to the fruit wall, whereas the fruit wall is undamaged. Occasionally, small fragments of the lemma and palea are still attached to the grain kernel. The small grain kernels, which are especially present in the lower and upper parts of the spike, are not fragmented.

The cleaning after fragmentation is usually done first by winnowing and/or sieving and then, as a final check, by hand-picking. Small quantities of diaspores of arable weeds and other impurities remain present in the food crop. Rachillate spikelets are easily recognized.

FIGURE 155

Bulgur

Obtained	Kastamonu (Kastamonu province, Turkey)
Origin	–
Date	August 2015
Collector(s)	R. T. J. Cappers
Context	Farmers market
GIA-accession	39621
Length/Diameter	–
Thickness	–
Weight	–
Quantity	–
Photographer	D. Fennema

Composition

Triticum monococcum ssp. *monococcum*	Einkorn	grain kernel

Impurities		
Apiaceae	Carrot family	fruit (mericarp)
Avena	Oat	fruit
Fabaceae	Pea family	seed
Galium	Bedstraw	fruit (mericarp)
cf. *Gypsophila*	Baby's breath	seed
Hordeum vulgare ssp. *vulgare*	Hulled 6-row Barley	floret
Triticum monococcum ssp. *monococcum*	Einkorn	chaff
Triticum monococcum ssp. *monococcum*	Einkorn	rachillate spikelet

Description

A coarse-ground bulgur made of unpolished grain kernels. Large quantities of einkorn bulgur are offered for sale at a farmers market. But this kind of bulgur is also in regular supply in shops in the city centre. This coarse bulgur is considered to be the best quality. The impurities consist of some rachillate spikelets and chaff of einkorn and diaspores of some arable weeds.

FIGURE 156

Bulgur

Obtained	Ihsangazi (Kastamonu province, Turkey)
Origin	–
Date	August 2015
Collector(s)	R. T. J. Cappers
Context	Farmhouse
GIA-accession	39638
Length/Diameter	–
Thickness	–
Weight	–
Quantity	–
Photographer	D. Fennema

Composition

Triticum monococcum ssp. *monococcum*	Einkorn	grain kernel

Impurities

Avena	Oat	floret
Avena	Oat	fruit
Avena fatua	Wild oat	floret
Claviceps	Ergot	sclerotium
Convolvulus arvensis	Field bindweed	seed
Elytrigia repens	Common couch	floret
Fabaceae	Pea family	seed
Galium	Bedstraw	fruit (mericarp)
cf. *Gypsophila*	Baby's breath	seed
Melampyrum arvense	Field cow-wheat	seed
Triticum monococcum ssp. *monococcum*	Einkorn	rachillate spikelet
–	stone	

Description

A coarse-ground bulgur made of unpolished grain kernels. The vegetative parts of the rachillate spikelets are removed between two large grinding stones that are powered by electricity. This is done in the barn of the farmhouse. Next, the cleaning is done, by means of an electric fan and sieves. The final check is done by hand-picking. This work is done by women. Small fragments of the chaff are still attached to the fruit wall. Impurities concern some rachillate spikelets of einkorn and diaspores of arable weeds. A conspicuous impurity are some fragments of the sclerotium of the fungus *Claviceps*, which may have been an infestation of the einkorn.

FIGURE 157

Bulgur

Obtained	Ihsangazi (Kastamonu province, Turkey)
Origin	–
Date	August 2015
Collector(s)	R. T. J. Cappers
Context	Private house
GIA-accession	39589
Length/Diameter	–
Thickness	–
Weight	–
Quantity	–
Photographer	D. Fennema

Composition

Triticum monococcum ssp. *monococcum*	Einkorn	grain kernel

Impurities

Ranunculus arvensis	Corn buttercup	fruit
Triticum monococcum ssp. *monococcum*	Einkorn	chaff

Description

Home-made bulgur that has been cleaned by winnowing (with an electric fan), sieving, and hand-picking. It is considered to be the best quality. The polished grain kernels are coarsely ground, through which most specimens have been divided lengthwise into two halves. Only a few small grain kernels are still complete. Impurities are limited and concern a single piece of chaff and a fruit of corn buttercup. A similar quality was obtained in Istanbul in February 2017 at a farmers market. The name for this product in Turkish is '*siyez bulgur*', which means 'Einkorn bulgur'. At this market, two qualities are offered for sale: uncleaned (10 TL/kg) and cleaned (15 TL/kg).

FIGURE 158

Bulgur

Obtained	Ihsangazi (Kastamonu province, Turkey)
Origin	–
Date	August 2015
Collector(s)	R. T. J. Cappers
Context	Private house
GIA-accession	39590
Length/Diameter	–
Thickness	–
Weight	–
Quantity	–
Photographer	D. Fennema

Composition

Triticum monococcum ssp. *monococcum*	Einkorn	grain kernel

Impurities

Allium	Onion	seed
Avena	Oat	fruit
Brassicaceae	Cabbage family	pedicel
Elytrigia repens	Common couch	floret
Galium	Bedstraw	fruit (mericarp)
Ranunculus arvensis	Corn buttercup	fruit
Triticum monococcum ssp. *monococcum*	Einkorn	chaff

Description

A medium-ground bulgur made of unpolished grain kernels. Most grain kernels have broken both lengthwise and several times crosswise. This bulgur is considered a lesser quality. The impurities concern a small quantity of einkorn chaff and diaspores of some arable weeds. Some pedicels of an unidentified member of the Cabbage family are also present. This bulgur was collected from a sheet on which bulgur is sun-dried.

FIGURE 159

Bulgur

Obtained	Ihsangazi (Kastamonu province, Turkey)
Origin	–
Date	August 2015
Collector(s)	R. T. J. Cappers
Context	Shop
GIA-accession	39616
Length/Diameter	–
Thickness	–
Weight	–
Quantity	–
Photographer	D. Fennema

Composition

Triticum monococcum ssp. *monococcum*	Einkorn	grain kernel

Impurities		
Avena	Oat	floret
Avena	Oat	fruit
Galium spp.	Bedstraw	fruit (mericarp)
Hordeum vulgare	Hulled Barley	floret
Lolium multiflorum	Italian rye-grass	floret
Triticum aestivum ssp. *aestivum*	Bread wheat	grain kernel
Triticum monococcum ssp. *monococcum*	Einkorn	chaff
Triticum monococcum ssp. *monococcum*	Einkorn	rachillate spikelet
cf. *Vaccaria hispanica*	Cowherb	seed

Description

A coarse-ground bulgur made of polished grain kernels. The polishing has resulted in the removal of only part of the bran. The germ has been removed in most of the kernels. The impurities consist of some chaff and rachillate spikelets of einkorn, grain kernels of bread wheat, florets of barley, and diaspores of arable weeds. This bulgur is export quality and is offered for sale pre-packaged in local shops as '*siyez bulgur*' ('Einkorn bulgur').

FIGURE 160

Bulgur

Obtained	Baneh (Kurdistan province, Iran)
Origin	–
Date	April 2017
Collector(s)	R. T. J. Cappers, M. Kaffashi & P. Hesami
Context	Drugstore
GIA-accession	40480
Length/Diameter	–
Thickness	–
Weight	–
Quantity	–
Photographer	D. Fennema

Composition

Triticum turgidum ssp. *dicoccon*	Emmer	grain kernel
Impurities		
Apiaceae	Carrot family	fruit (mericarp)
Avena	Oat	fruit
Cichorium intybus	Chicory	fruit
Elytrigia repens	Common couch	floret
Galium	Bedstraw	fruit (mericarp)
Lens culinaris	Lentil	seed
Persicaria	Knotweed	fruit
Rumex	Docks	seed
Scandix pecten-veneris	Shepherd's needle	fruit (mericarp)
Triticum turgidum ssp. *dicoccon*	Emmer	rachillate spikelet
–	stone	

Description

The rachillate spikelets have been par-boiled, after which the vegetative parts have been removed from the grain kernels (de-graining). This was most probably done by grinding. The grain kernels are neither polished nor ground. The minor amount of fragmentation that is present can be attributed to the de-graining. The impurities consist of rachillate spikelets of emmer, lentil, diaspores of some arable weeds, and stones.

FIGURE 161

Bulgur

Obtained	Sanandaj (Kurdistan province, Iran)
Origin	–
Date	April 2017
Collector(s)	R. T. J. Cappers & M. Kaffashi
Context	Bazaar
GIA-accession	40366
Length/Diameter	–
Thickness	–
Weight	–
Quantity	–
Photographer	D. Fennema

Composition

Triticum turgidum ssp. *dicoccon*	Emmer	grain kernel
Impurities		
Aegilops	Goat beard	fruit
Apiaceae	Carrot family	fruit (mericarp)
Centaurea	Knapweed	fruit
Chenopodiaceae	Goosefoot family	fruit
Cichorium intybus	Chicory	fruit
Galium	Bedstraws	fruit (mericarp)
Haplophyllum	–	fruit
Hordeum vulgare ssp. *vulgare*	Hulled 6-row Barley	floret
Lens culinaris	Lentil	seed
Linum usitatissimum	Flax	seed
Lupinus	Lupin	seed
Poa Meadow grass	flowering spikelet	
Salvia sclarea	Clary	fruit (mericarp)
Triticum turgidum ssp. *dicoccon*	Emmer	chaff
Triticum turgidum ssp. *dicoccon*	Emmer	culm
Triticum turgidum ssp. *dicoccon*	Emmer	rachillate spikelet
Triticum turgidum ssp. *dicoccon*	Emmer	root
Triticum turgidum ssp. *durum*	Hard wheat	grain kernel
Triticum turgidum ssp. *durum*	Hard wheat	rachis fragment
–	stone	

Description

A medium-ground bulgur made of unpolished grain kernels. Most fragments concern a quarter of the grain kernel, thus fragmented both lengthwise and crosswise. A small proportion of the grains are still whole or only fragmented once (usually lengthwise). Whole rachillate spikelets as well as the typical fragments of the by-product of de-graining are present in a reasonable number: rachis fragments with the basal parts of the glumes, in some cases divided into two parts, as well as rachis fragments without the remains of the glumes. Apical rachillate spikelets with a twisted spikelet are present and support a positive subspecies identification (and exclude einkorn). The impurities consist of remains of barley, hard wheat, lentil, and flax; diaspores of arable weeds; and small stones. The number of flax seeds and stone fragments is conspicuous.

FIGURE 162

Bulgur

Obtained	Baneh (Kurdistan province, Iran)
Origin	–
Date	April 2017
Collector(s)	R. T. J. Cappers, M. Kaffashi & P. Hesami
Context	Drugstore
GIA-accession	40528
Length/Diameter	–
Thickness	–
Weight	–
Quantity	–
Photographer	D. Fennema

Composition

Triticum turgidum ssp. *dicoccon*	Emmer	grain kernel

Impurities		
Ammi	–	fruit (mericarp)
Anethum graveolens	Dill	fruit (mericarp)
Avena	Oat	floret
Avena	Oat	fruit
Bupleurum lancifolium	–	fruit (mericarp)
Cephalaria syriaca	–	fruit and perianth
Cerastium	Mouse ear	seed
Cichorium intybus	Chicory	fruit
Consolida	Larkspur	seed
Convolvulus arvensis	Field bindweed	seed
Coriandrum sativum	Coriander	fruit (mericarp)
Asteraceae	Daisy family	fruit
Echium	Viper's buglossus	fruit (mericarp)
Fabaceae	Pea family	seed
Galium	Bedstraw	fruit (mericarp)
Hordeum vulgare	Hulled Barley	floret
Lens culinaris	Lentil	seed
Lolium temulentum	Darnel	floret
Neslia paniculata	Ball mustard	fruit
Nigella sativa	Black cumin	seed
Persicaria	Knotweed	fruit and perianth
Phaseolus vulgaris	Field bean	seed
Poa	Meadow grass	flowering spikelet
Poaceae	Grass family	fruit
Rumex acetosella	Sheep's sorrel	fruit
Scandic pecten-veneris	Shepherd's needle	fruit (mericarp)
Secale cereale	Rye	grain kernel
Sorghum halepense	Johnson grass	spikelet
Triticum turgidum ssp. *dicoccon*	Emmer	chaff
Triticum turgidum ssp. *dicoccon*	Emmer	culm
Triticum turgidum ssp. *dicoccon*	Emmer	rachillate spikelet
Triticum turgidum ssp. *dicoccon*	Emmer	root
–	stone	

Description

A bulgur made of unpolished whole grain kernels. A small fraction has fragmented into different size qualities during de-graining. The impurities consist of the vegetative parts of emmer, plant parts of economic plants (including field bean, lentil, barley, dill, coriander, black cumin, and rye), and diaspores of arable weeds. The seeds and fruits of the economic plants most probably became mixed in with the bulgur in the shop where the bulgur was offered for sale.

FIGURE 163

Bulgur

Obtained	Ilam (Ilam province, Iran)
Origin	–
Date	June 2017
Collector(s)	M. Kaffashi
Context	Shop
GIA-accession	40832
Length/Diameter	–
Thickness	–
Weight	–
Quantity	–
Photographer	D. Fennema

Composition

Triticum turgidum ssp. *dicoccon*	Emmer	grain kernel

Impurities		
Anchusa	Alkanet	fruit (mericarp)
Apiaceae spp.	Carrot family	fruit (mericarp)
Brassicaceae spp.	Cabbage family	pedicel
Bromus	Brome	fruit
Cannabis sativa	Hemp	fruit
Cephalaria syriaca	–	fruit and perianth
Cichorium intybus	Chicory	fruit
Convolvulus arvensis	Field bindweed	seed
Coriandrum sativum	Coriander	fruit (schizocarp)
Cuminum cyminum	Cumin	fruit (schizocarp)
Fallopia convolvulus	Black bindweed	fruit
Hordeum vulgare ssp. *vulgare*	Hulled 6-row Barley	floret
Lallemantia spp.	–	fruit (mericarp)
Lamiaceae	Dead-nettle family	fruit (mericarp)
Neslia paniculata	Ball mustard	fruit
Panicum miliaceum	Common millet	floret
Pistachia terebinthus	Turpentine tree	fruit
Poa Meadow grass	floret	
Sanguisorba	Burnet	fruit with exocarp
Scandic pecten-veneris	Shepherd's needle	fruit (mericarp)
Scandic stellata	–	fruit (mericarp)
Sesamum indicum	Sesame	seed
Triticum aestivum ssp. *aestivum*	Bread wheat	grain kernel
Triticum aestivum ssp. *aestivum*	Bread wheat	rachis fragment
Triticum turgidum ssp. *dicoccon*	Emmer	chaff
Triticum turgidum ssp. *dicoccon*	Emmer	rachillate spikelet
Triticum turgidum ssp. *dicoccon*	Emmer	root
Vigna radiata	Mung bean	seed
–	insect	
–	stone	

Description

A medium-ground bulgur made of unpolished grain kernels. Most of the grain kernels have fragmented lengthwise and crosswise, but intact specimens and those fragmented only lengthwise are also present. The impurities are present in a considerable quantity. In addition to some whole rachillate spikelets, including the diagnostic top specimens with their twisted spikelet, many fragments of chaff and culm are also present. The abundance of both fragments with root scars and stones suggest that at least part of the emmer had become uprooted during the harvest. The presence of a variety of economic plants indicates that the bulgur has also become mixed during storage. They include hemp, coriander, cumin, barley, common millet, turpentine tree, sesame, bread wheat, and mung bean. In addition, diaspores of quite a number of arable weeds are present. One beetle is present as well.

FIGURE 164

Bulgur

Obtained	Tabriz (East Azerbaijan province, Iran)
Origin	–
Date	July 2017
Collector(s)	R. T. J. Cappers & M. Kaffashi
Context	Bazaar
GIA-accession	40812
Length/Diameter	–
Thickness	–
Weight	–
Quantity	–
Photographer	D. Fennema

Composition

Triticum turgidum ssp. *dicoccon*	Emmer	grain kernel

Impurities		
Apiaceae	Carrot family	fruit (mericarp)
Avena	Oat	floret
Brassicaceae spp.	Cabbage family	fruit
Brassicaceae	Cabbage family	seed
Convolvulus	Field bindweed	seed
Elytrigia repens	Common couch	floret
Galium	Bedstraw	fruit (mericarp)
Hordeum vulgare ssp. *vulgare*	Hulled 6-row Barley	floret
Malva	Mallow	seed
Medicago	Medick	seed
Persicaria	Knotweed	fruit and perianth
Poaceae	Grass family	fruit
Setaria	Bristle grass	floret
Sorghum halepense	Johnson grass	spikelet
Triticum aestivum ssp. *aestivum*	Bread wheat	grain kernel
Triticum turgidum ssp. *dicoccon*	Emmer	chaff
Triticum turgidum ssp. *dicoccon*	Emmer	rachillate spikelet
unidentified	–	culm
–	stone	

Description

A coarse-ground bulgur made of unpolished grain kernels. Fragments of the bran are still attached to the grain kernel. During separation of the grain kernels from the vegetative parts, a number of embryos have been removed from the grain kernels as well, and these are also present in the sample. Two landraces of barley are represented: one with straw-coloured chaff and one with dark grey-coloured chaff. Impurities consist of economic plants, including barley and bread wheat; diaspores of arable weeds; culm fragments; and stones.

FIGURE 165

Bulgur

Obtained	Tabriz (East Azerbaijan province, Iran)
Origin	–
Date	July 2017
Collector(s)	R.T.J. Cappers & M. Kaffashi
Context	Bazaar
GIA-accession	40809
Length/Diameter	–
Thickness	–
Weight	–
Quantity	–
Photographer	D. Fennema

Composition

Triticum turgidum ssp. *dicoccon*	Emmer	grain kernel

Impurities		
Brassicaceae spp.	Cabbage family	seed
Capsella bursa-pastoris	Shepherd's purse	seed
Cephalaria syriaca	–	fruit and perianth
Chenopodium	Goosefoot	fruit
Cichorium intybus	Chicory	fruit
Convolvulus	Field bindweed	seed
cf. *Gypsophila*	Baby's breath	seed
Hordeum vulgare ssp. *vulgare*	Hulled 6-row Barley	floret
Lallemantia	–	fruit (mericarp)
Lamiaceae	Dead-nettle family	fruit (mericarp)
Persicaria spp.	Knotweed	fruit
Silene spp.	Campion	seed
Triticum turgidum ssp. *dicoccon*	Emmer	chaff
Triticum turgidum ssp. *dicoccon*	Emmer	culm
Triticum turgidum ssp. *dicoccon*	Emmer	rachillate spikelet
–	stone	

Description

A coarse-ground bulgur made of polished grain kernels. Small fragments of the bran are still attached to the grain kernel. The impurities consist of vegetative plant parts of emmer, florets of barley, diaspores of arable weeds, and stones.

FIGURE 166

Bulgur

Obtained	Düzce (Düzce province, Turkey)
Origin	Gaziantep (Gaziantep province, Turkey)
Date	February 2017
Collector(s)	R. T. J. Cappers, H. Balcı & L. Kutlu
Context	Farmers market
GIA-accession	40287
Length/Diameter	–
Thickness	–
Weight	–
Quantity	–
Photographer	D. Fennema

Composition

Triticum turgidum ssp. *durum*	Hard wheat	grain kernel

Impurities

Aegilops tauschii	Goat beard	rachillate spikelet
Avena	Oat	fruit
Hordeum vulgare	Hulled Barley	floret

Description

Home-made bulgur, which is designated as 'village bulgur'. It is a coarse-ground bulgur made of polished grain kernels.

FIGURE 167

Bulgur

Obtained	Altunhisar (Nigde province, Turkey)
Origin	–
Date	August 2014
Collector(s)	F. Fantone
Context	–
GIA-accession	34973
Length/Diameter	–
Thickness	–
Weight	–
Quantity	–
Photographer	D. Fennema

Composition

Triticum turgidum ssp. *durum*	Hard wheat	grain kernel

Impurities

Hordeum vulgare	Hulled Barley	floret

Description

A coarse-ground bulgur made of polished grain kernels. Small fragments of the bran are still present on the fruit wall. The larger grain kernels have broken into two or three fragments, and the smaller ones are still complete. It is striking that most of the grain kernels have broken crosswise rather than longitudinally. The Turkish word for this bulgur, being '*iri pilavlik*', which literally means 'coarse-ground rice', is only partly correct, because the designation '*pilavlik*' is usually applied to grain kernels that are predominantly split lengthwise. The grain kernels that are broken lengthwise are somewhat longer, and this suggests that the grain was separated into different size classes before grinding.

FIGURE 168

Bulgur

Obtained	Kastamonu (Kastamonu province, Turkey)
Origin	–
Date	August 2015
Collector(s)	R.T.J. Cappers
Context	Farmers market
GIA-accession	39620
Length/Diameter	–
Thickness	–
Weight	–
Quantity	–
Photographer	D. Fennema

Composition

Triticum turgidum ssp. *durum*	Hard wheat	grain kernel
Impurities		
Hordeum vulgare	Hulled Barley	floret
Lens culinaris	Lentil	seed
–	insect	

Description

The grain kernels have been polished by pounding, probably with a wooden hammer in a stone mortar. The polished grain kernels have been par-boiled but not ground. Although the grain kernels are still whole, they have many irregular cracks, giving their surface a craquelée effect. The fragmented grain kernels, which constitute only a small proportion, are the result of the polishing.

The bulgur has become infested after preparation with two kinds of insects: a saw-toothed grain beetle and a stored-product moth.

FIGURE 169

Bulgur

Obtained	Karaman (Karaman province, Turkey)
Origin	–
Date	August 2016
Collector(s)	Y. Okur
Context	Market
GIA-accession	39885
Length/Diameter	–
Thickness	–
Weight	–
Quantity	–
Photographer	D. Fennema

Composition

Cicer arietinum	Chick pea	seed
Triticum turgidum ssp. *durum*	Hard wheat	grain kernel

Description

A coarse-ground bulgur made of polished grain kernels mixed with whole seeds of chick pea. The seeds of the chick pea have also been boiled, resulting in the bursting of the seedcoat. The name for this product in Turkish is '*nohutlu bulgur*', which means 'chick pea bulgur'. The remaining cooking time is 15 minutes. The shelf life is 1 year according to the packaging.

Nutritional value (100 g)

carbohydrates	62.8 g
lipids	3.1 g
proteins	12.9 g

FIGURE 170

Bulgur

Obtained	Karaman (Karaman province, Turkey)
Origin	–
Date	August 2016
Collector(s)	Y. Okur
Context	Market
GIA-accession	39887
Length/Diameter	–
Thickness	–
Weight	–
Quantity	–
Photographer	D. Fennema

Composition

Lens culinaris	Lentil	seed
Triticum turgidum ssp. *durum*	Hard wheat	grain kernel

Description
A coarse-ground bulgur made of polished grain kernels mixed with whole seeds of green lentil. The seeds of the lentil have also been boiled, resulting in the bursting of the seedcoat along the edge. The name for this product in Turkish is '*mercimekli bulgur*', which means 'lentil bulgur'. The remaining cooking time is 15 minutes. The shelf life is 1 year according to the packaging.

Nutritional value (100 g)

carbohydrates	62.3 g
lipids	1.7 g
proteins	14.2 g

FIGURE 171

Bulgur

Obtained	Derik (Al-Hasakah governorate, Syria)
Origin	–
Date	October 2014
Collector(s)	D. Alshawish
Context	–
GIA-accession	35093
Length/Diameter	–
Thickness	–
Weight	–
Quantity	–
Photographer	D. Fennema

Composition

Triticum turgidum ssp. *durum*	Hard wheat	grain kernel

Impurities

Fabaceae	Pea family	seedcoat
cf. *Vaccaria hispanica*	Cowherb	seed
–	stone	

Description

A fine-ground bulgur made of polished grain kernels. Small parts of the bran are still present at the hilum groove. The grain kernels are soaked prior to boiling. After par-boiling, the grain is first polished and then ground. The name in Kurdish for this bulgur is '*savar*'. Impurities are limited and consist of fragments of arable weeds and stones.

FIGURE 172

Bulgur

Obtained	Tabriz (East Azerbaijan province, Iran)
Origin	–
Date	June 2017
Collector(s)	R. T. J. Cappers & M. Kaffashi
Context	Bazaar
GIA-accession	40947
Length/Diameter	–
Thickness	–
Weight	–
Quantity	–
Photographer	D. Fennema

Composition

Triticum turgidum ssp. *durum*	Hard wheat	grain kernel
Impurities		
Apiaceae spp.	Carrot family	fruit (mericarp)
Apiaceae	Carrot family	pedicel
Asteraceae	Daisy family	fruit
Bromus	Brome	fruit
Cephalaria syriaca	–	fruit and perianth
Convolvulus arvensis	Field bindweed	seed
Elytrigia repens	Common couch	floret
Galium	Bedstraw	fruit (mericarp)
Hordeum vulgare ssp. *vulgare*	Hulled 6-row Barley	floret
Secale cereale	Rye	grain kernel
unidentified	–	culm fragments
–	dung	
–	stone	

Description

A medium-ground bulgur made of unpolished grain kernels. A small number of grain kernels are still complete or only split lengthwise. The impurities consist of some economic plants, including barley and rye; diaspores of arable weeds; stones; and a small dung dropping.

FIGURE 173

Bulgur

Obtained	Ardabil (Ardabil province, Iran)
Origin	–
Date	July 2017
Collector(s)	R. T. J. Cappers & M. Kaffashi
Context	Bazaar
GIA-accession	40866
Length/Diameter	–
Thickness	–
Weight	–
Quantity	–
Photographer	D. Fennema

Composition

Triticum turgidum ssp. *durum*	Hard wheat	grain kernel
Impurities		
Agrostemma githago	Corncockle	seed
Apiaceae	Carrot family	fruit (mericarp)
Carum carvi	Caraway	fruit (mericarp)
Convolvulus	Field bindweed	seed
Echium	Viper's buglossus	fruit (mericarp)
Elytrigia repens	Common couch	floret
Fabaceae	Pea family	seed
Galium	Bedstraw	fruit (mericarp)
Hordeum vulgare ssp. *vulgare*	Hulled 6-row Barley	floret
Lathyrus sativus	Grass pea	seed
Neslia paniculata	Ball mustard	fruit
Persicaria	Knotweed	fruit
cf. *Ranunculus arvensis*	Corn buttercup	fruit
Scandix	Shepherd's needle	fruit (mericarp)
Secale cereale	Rye	grain kernel
unidentified	–	culm fragment
–	stone	

Description

A medium-ground bulgur made of polished grain kernels. Fragments of the bran are still present on many of the grain kernel fragments. A small number of grain kernels is not par-boiled. The impurities consist of seeds and fruits of some economic plants, including grass pea, caraway, barley, and rye; diaspores of arable weeds; and stones.

FIGURE 174

Bulgur

Obtained	Hamedan (Hamedan province, Iran)
Origin	Malayer (Hamedan province, Iran)
Date	July 2017
Collector(s)	R.T.J. Cappers & M. Kaffashi
Context	Bazaar
GIA-accession	40797
Length/Diameter	–
Thickness	–
Weight	–
Quantity	–
Photographer	D. Fennema

Composition

Triticum turgidum ssp. *durum*	Hard wheat	grain kernel
Impurities		
Capsella bursa-pastoris	Shepherd's purse	seed
Caryophyllaceae	Pink family	seed
Medicago	Medick	seed
Silene	Campion	seed
Triticum turgidum ssp. *durum*	Hard wheat	grain kernel
unidentified	–	culm

Description

A medium-ground bulgur made of polished grain kernels. The smaller among the grain kernels remain unfragmented. Small fragments of the bran are still attached to the grain kernels. This quality is used for making tarhana. The impurities consist of hard wheat, diaspores of arable weeds, a culm fragment, and stones.

FIGURE 175

Bulgur

Obtained	Erbil (Erbil governorate, Iraq)
Origin	–
Date	June 2014
Collector(s)	R. T. J. Cappers, F. Fantone & S. Filatova
Context	Shop
GIA-accession	34795
Length/Diameter	–
Thickness	–
Weight	–
Quantity	–
Photographer	D. Fennema

Composition

Triticum turgidum ssp. *durum*	Hard wheat	grain kernel
Impurities		
Lolium temulentum	Darnel	floret
Triticum aestivum ssp. *aestivum*	Bread wheat	grain kernel

Description

A medium-ground bulgur made of polished grain kernels. Most fragments consist of a quarter of the grain kernel. Smaller specimens are present as half grain kernels, mostly split lengthwise, or even as whole grain kernels. Impurities concern some grain kernels of bread wheat and some partly de-grained florets of darnel.

FIGURE 176

Bulgur

Obtained	Erbil (Erbil governorate, Iraq)
Origin	–
Date	June 2014
Collector(s)	R. T. J. Cappers, F. Fantone & S. Filatova
Context	Shop
GIA-accession	34796
Length/Diameter	–
Thickness	–
Weight	–
Quantity	–
Photographer	D. Fennema

Composition

Triticum turgidum ssp. *durum*	Hard wheat	grain kernel

Impurities

Hordeum vulgare	Hulled Barley	floret
Poaceae	Grass family	fruit
–	stone	

Description

A medium-ground bulgur made of polished grain kernels. A few fragments remain unpolished. Small bran fragments are also present in the hilum groove. The impurities concern a fragment of a pearled barley grain kernel, a fragment of an unidentified wild grass, and a small stone.

FIGURE 177

Bulgur

Obtained	Boskin (Al-Sulaimaniyah governorate, Iraq)
Origin	–
Date	June 2014
Collector(s)	R.T.J. Cappers, F. Fantone & S. Filatova
Context	Shop
GIA-accession	34845
Length/Diameter	–
Thickness	–
Weight	–
Quantity	–
Photographer	D. Fennema

Composition

Triticum turgidum ssp. *durum*	Hard wheat	grain kernel

Impurities

–	stone

Description

A medium-ground bulgur made of polished grain kernels. Small fragments of the bran are present, especially in the hilum groove. The only impurity consists of some small stones.

FIGURE 178

Bulgur

Obtained	Erbil (Erbil governorate, Iraq)
Origin	–
Date	June 2014
Collector(s)	R. T. J. Cappers, F. Fantone & S. Filatova
Context	Shop
GIA-accession	34797
Length/Diameter	–
Thickness	–
Weight	–
Quantity	–
Photographer	D. Fennema

Composition

Triticum turgidum ssp. *durum*	Hard wheat	grain kernel

Impurities

–	insect

Description

A find-ground bulgur made of polished grain kernels. Bran fragments are few and originate from the hilum groove. The presence of a relatively large larva is indicative of insect infestation during storage.

FIGURE 179

Bulgur

Obtained	Boskin (Al-Sulaimaniyah governorate, Iraq)
Origin	–
Date	June 2014
Collector(s)	R. T. J. Cappers, F. Fantone & S. Filatova
Context	Shop
GIA-accession	34800
Length/Diameter	–
Thickness	–
Weight	–
Quantity	–
Photographer	D. Fennema

Composition

Triticum turgidum ssp. *durum*	Hard wheat	grain kernel

Impurities

Hordeum vulgare	Hulled Barley	floret

Description

A find-ground bulgur made of unpolished grain kernels. Small fragments of the bran are present. The shelf life is 2 years according to the packaging.

Nutritional value (100 g)

carbohydrates	67.8 g
lipids	1.5 g
proteins	12.5 g

FIGURE 180

Bulgur

Obtained	Baneh (Kurdistan province, Iran)
Origin	–
Date	April 2017
Collector(s)	R. T. J. Cappers, M. Kaffashi & P. Hesami
Context	Drugstore
GIA-accession	40561
Length/Diameter	–
Thickness	–
Weight	–
Quantity	–
Photographer	D. Fennema

Composition

Triticum turgidum ssp. *durum*	Hard wheat	grain kernel
Impurities		
Apiaceae spp.	Carrot family	fruit (mericarp)
Fabaceae spp.	Pea family	seed
Galium	Bedstraws	fruit (mericarp)
Hordeum vulgare	Hulled Barley	floret
Lens culinaris	Lentil	seed
Persicaria	Knotweed	fruit
Poaceae	Grass family	fruit
Triticum turgidum ssp. *durum*	Hard wheat	culm
Triticum turgidum ssp. *durum*	Hard wheat	rachis fragment
–	stone	

Description

A medium-ground bulgur made of unpolished grain kernels. A small quantity of diaspores of arable weeds is present, and some of these are polished and broken as well.

FIGURE 181

Bulgur with noodles

Obtained	Erbil (Erbil governorate, Iraq)
Origin	–
Date	June 2014
Collector(s)	R.T.J. Cappers, F. Fantone & S. Filatova
Context	Shop
GIA-accession	34844
Length/Diameter	–
Thickness	–
Weight	–
Quantity	–
Photographer	D. Fennema

Composition

Triticum turgidum ssp. *durum*	Hard wheat	grain kernel

Description

The find-ground bulgur is mixed with pasta (shaped into fine noodles) and thus represents a mixture of two different processes of possibly the same cereal. It is assumed that the pasta has been boiled as well to make it possible that both ingredients of the mixture can be cooked to the point of doneness with the same final cooking time. The grain kernels have been polished before grinding, and small fragments of the bran are only present in the hilum groove.

FIGURE 182

Bulgur

Obtained	Rabat (Rabat-Salé-Kénitra region, Morocco)
Origin	–
Date	March 2016
Collector(s)	R.T.J. Cappers
Context	Bazaar
GIA-accession	39805
Length/Diameter	–
Thickness	–
Weight	–
Quantity	–
Photographer	D. Fennema

Composition

Triticum turgidum ssp. *durum*	Hard wheat	grain kernel

Impurities

Avena	Oat	fruit
Avena	Oat	floret
Avena fatua	Wild oat	floret
Lens culinaris	Lentil	seed
Lolium temulentum	Darnel	floret

Description

A coarse-ground bulgur made of unpolished grain kernels. The grain kernels vary in size and shape from short to more elongate. Impurities concern some seeds of lentil and a few diaspores of arable weeds.

FIGURE 183

Bulgur

Obtained	Cairo (Cairo governorate, Egypt)
Origin	Beni Suef (Beni Suef governorate, Egypt)
Date	August 2017
Collector(s)	A. Shahat
Context	Shop
GIA-accession	40990
Length/Diameter	–
Thickness	–
Weight	–
Quantity	–
Photographer	D. Fennema

Composition

Triticum turgidum ssp. *durum*	Hard wheat	grain kernel
Triticum aestivum ssp. *aestivum*	Bread wheat	grain kernel

Impurities		
Avena	Oat	awn
Avena	Oat	chaff
Avena	Oat	fruit
Malva nicaeensis/parviflora	French/Least mallow	fruit (mericarp)
Melilotus indicus	Small melilot	seed
Phalaris	Canary-grass	floret
–	stone	

Description

This bulgur is a mixture of hard wheat and bread wheat. It seems as if the hard wheat has been polished before grinding, whereas the bread wheat has not been polished. It is also striking that the polished grain is ground into small fragments, whereas the bread wheat is represented by coarse fragments and even a considerable number of intact grain kernels. The impurities consist of diaspores of some arable weeds and stones. The awns of oat are present as small units.

FIGURE 184

Bulgur

Obtained	Göcer (Karaman province, Turkey)
Origin	–
Date	August 2014
Collector(s)	Y. Okur
Context	–
GIA-accession	35099
Length/Diameter	–
Thickness	–
Weight	–
Quantity	–
Photographer	D. Fennema

Composition

Triticum cf. *aestivum* ssp. *aestivum*	Cf. Bread wheat	grain kernel

Impurities

Elytrigia repens	Common couch	floret
Poaceae	Grass family	fruit
Triticum cf. *aestivum* ssp. *aestivum*	Cf. Bread wheat	culm fragment

Description

A medium-ground bulgur made of unpolished grain kernels. The grain kernels are first washed. Then they are dried and sieved to remove the disintegrated clay particles. Next, the grain is par-boiled and dried again. Finally, the grain is ground.

FIGURE 185

Bulgur

Obtained	Istanbul (Istanbul province, Turkey)
Origin	Gaziantep (Gaziantep province, Turkey)
Date	September 2014
Collector(s)	R. T. J. Cappers & Y. Okur
Context	Shop
GIA-accession	35010
Length/Diameter	–
Thickness	–
Weight	–
Quantity	–
Photographer	D. Fennema

Composition

Triticum aestivum ssp. *aestivum*	Bread wheat	grain kernel
Impurities		
Avena	Oat	fruit
Convolvulus	Field bindweed	seed
Galium	Bedstraws	fruit (mericarp)
Hordeum vulgare	Hulled Barley	floret
unidentified	–	–
Triticum turgidum ssp. *durum*	Hard wheat	grain kernel

Description

A coarse-ground bulgur made of polished grain kernels. This bulgur is characterized by a reddish colour, which is referred to in the name in Turkish, which is '*esmer pilavlik bulgur*', whereby '*esmer*' designates the reddish bread wheat and '*pilavlik*' means rice, here referring to the coarse grinding. Due to the deep ventral groove, the wheat grain kernel easily fragments in a longitudinal direction, resulting in a shape that resembles that of a whole grain kernel of rice (*Oryza sativa*). The grain kernels in a spikelet of naked wheat vary in size, gradually decreasing in size from the bottom position to the top position. The small-sized grain kernels escape the grinding.

FIGURE 186

Bulgur

Obtained	Derik (Al-Hasakah governorate, Syria)
Origin	–
Date	October 2014
Collector(s)	D. Alshawish
Context	–
GIA-accession	35092
Length/Diameter	–
Thickness	–
Weight	–
Quantity	–
Photographer	D. Fennema

Composition

Triticum aestivum ssp. *aestivum*	Bread wheat	grain kernel
Impurities		
Convolvulus arvensis	Field bindweed	seed
Hordeum vulgare ssp. *vulgare*	Hulled 6-row Barley	floret
Lens culinaris	Lentil	seed
Triticum aestivum ssp. *aestivum*	Bread wheat	chaff

Description

The grain kernels have been soaked in water prior to being par-boiled. This is an intermediate stage in the processing of the bulgur (Kurdish name: *genime kelandi* (or: *gekelandi*). The next steps will include polishing and grinding. The grain kernels are somewhat shrunken. Also present are large bran fragments that are almost complete. Because the germ has not increased in size, it can be deduced that the soaking period was short. The impurities consist of some chaff fragments of the wheat, lentil, and barley and a diaspore of an arable weed. The florets of barley are either straw-coloured or grey-brown and thus represent two landraces.

FIGURE 187

Bulgur

Obtained	Urmia (East Azerbaijan province, Iran)
Origin	–
Date	August 2017
Collector(s)	J. Abdeli
Context	–
GIA-accession	41006
Length/Diameter	–
Thickness	–
Weight	–
Quantity	–
Photographer	D. Fennema

Composition

Triticum aestivum ssp. aestivum	Bread wheat	grain kernel

Impurities

Asteraceae	Daisy family	fruit
Convolvulus arvensis	Field bindweed	seed
Galium	Bedstraw	fruit (mericarp)
Hordeum vulgare	Hulled Barley	floret
Scandix pecten-veneris	Shepherd's needle	fruit (mericarp)
Triticum aestivum ssp. aestivum	Bread wheat	chaff
Triticum turgidum ssp. durum	Hard wheat	grain kernel

Description

A coarse-ground bulgur made of unpolished grain kernels. The grain kernels are first cleaned by hand-picking. Then the grain is washed. It is said that this is especially done to clean the grain from any evidence of visits by mice. Next the grain is par-boiled, after which it is drained under running water and then dried in the sun. The grain kernels have been ground with a rotary quern by. H. Vahidi. Smaller fragments are not sieved out. According to the collector, this bulgur is considered the best Iranian quality. The white colour, being typical of bread wheat, is appreciated above the yellow colour (being typical of hard wheat). The impurities concern pearled florets of barley and a few diaspores of arable weeds.

FIGURE 188

Bulgur

Obtained	Tehran (Tehran province, Iran)
Origin	Isfahan (Isfahan province, Iran)
Date	April 2017
Collector(s)	R. T. J. Cappers & M. Kaffashi
Context	Shop
GIA-accession	40405
Length/Diameter	–
Thickness	–
Weight	–
Quantity	–
Photographer	D. Fennema

Composition

Triticum aestivum ssp. *aestivum*	Bread wheat	grain kernel

Impurities		
Carum carvi	Caraway	fruit (mericarp)
Carum carvi	Caraway	pedicel
Galium	Bedstraws	fruit (mericarp)
Poaceae	Grass family	spikelet
Rumex	Docks	fruit
Triticum aestivum ssp. *aestivum*	Bread wheat	culm fragment
–	stone	

Description

A coarse-ground bulgur made of unpolished grain kernels. A small quantity of fruits of caraway is present. Some fruits are damaged, and it is therefore considered as an impurity rather than an additive to improve the taste.

FIGURE 189

12
Pasta

Pasta is made of unleavened dough. The dough is made from a mixture of meal or flour, water, and salt. The mixture is thus similar to that which is used for bread making. The main differences between pasta and unleavened bread concern the size and shape of the units and the method of heating.

Modern pasta is made from pulverized wheat grain, and of Hard wheat (*Triticum turgidum* ssp. *durum*) grain in particular. When fragmentation by pounding or grinding results in a mixture of granules and powdery fragments, it is the latter that are sieved out to be used for pasta production.

Traditionally, pasta is produced in a wide variety of shapes. Shaping is done by flattening the dough and then cutting it with a knife or by pressing the dough through a screen. The shaped pasta can be stored in that state or after being dry-heated or wet-heated for a short time and then dried. Partial heating has the advantage that eggs of insects and mites are killed, thereby improving the shelf life. Scorch marks present on homemade pasta are indicative of dry-heating. The final heating of pasta is done by boiling. The availability of pottery could thus have facilitated the production of this kind of cereal food.

Pasta

Obtained	Kastamonu (Kastamonu province, Turkey)
Origin	–
Date	August 2015
Collector(s)	R. T. J. Cappers
Context	Farmers market
GIA-accession	39637
Length/Diameter	–
Thickness	–
Weight	–
Quantity	–
Photographer	D. Fennema

Composition

Triticum	Wheat	grain kernel
–	salt	
–	water	

Description

This pasta is home-made. The absence of scorch marks suggests that the flattened dough was not dry heated prior to shaping. Concentrations of flour are present at the surface. This pasta is designated as '*kare eriste*' ('square pasta'). The squares of this pasta are rather large. A similar kind of pasta, but shaped into smaller units, obtained at a farmers market in Istanbul originated from a small village near Inebolu (Kastamonu province).

FIGURE 190

Pasta

Obtained	Istanbul (Istanbul province, Turkey)
Origin	–
Date	September 2014
Collector(s)	R. T. J. Cappers & Y. Okur
Context	Shop
GIA-accession	34989
Length/Diameter	–
Thickness	–
Weight	–
Quantity	–
Photographer	D. Fennema

Composition

Triticum	Wheat	grain kernel
–	salt	
–	water	

Description

This pasta is possibly home-made. The square units vary in thickness and size. Some pieces have a curved profile because they originate from the edge of the flattened dough ball. Scorch marks are absent, indicating that the flattened dough was not dry heated prior to shaping. Concentrations of flour are present at the surface. This pasta is designated as '*kare eriste*' ('square pasta'). The squares of this pasta are rather small.

FIGURE 191

12 Pasta

Pasta

Obtained	Cakirbag (Karaman province, Turkey)
Origin	–
Date	August 2014
Collector(s)	Y. Okur
Context	Private house
GIA-accession	34969
Length/Diameter	–
Thickness	–
Weight	–
Quantity	–
Photographer	D. Fennema

Composition

Triticum turgidum ssp. *durum*	Hard wheat	Grain kernel
–	egg	
–	salt	
–	water	

Description

This pasta is home-made. The pasta is designated as '*erişte macaroni*' ('vermicelli pasta'), referring to the elongated shape. After mixing the meal with egg, salt, and water, the resulting dough is shaped into small balls. After proving for some time, the balls are shaped into flat, round loaves by using a rolling pin, on a wooden table. The thin loaves are heated briefly on a hot metal plate and then stacked on top of each other. A pile of thin loaves is cut into elongated units. The pasta can be stored for half a year. The short heating will kill any eggs of insects and mites that might be present in the meal and thus reduces the chance of the pasta spoiling.

FIGURE 192

Pasta

Obtained	Kastamonu (Kastamonu province, Turkey)
Origin	–
Date	August 2015
Collector(s)	R. T. J. Cappers
Context	Farmers market
GIA-accession	39635
Length/Diameter	–
Thickness	–
Weight	–
Quantity	–
Photographer	D. Fennema

Composition

Triticum	Wheat	grain kernel
–	salt	
–	water	

Description

This pasta is home-made. The pasta does not show any scorch marks. The pasta may have been par-boiled after shaping.

FIGURE 193

Pasta

Obtained	Hendek (Sakarya province, Turkey)
Origin	Gerede (Bolu province, Turkey)
Date	February 2017
Collector(s)	R. T. J. Cappers & H. Balcı
Context	Shop
GIA-accession	40303
Length/Diameter	–
Thickness	–
Weight	–
Quantity	–
Photographer	D. Fennema

Composition

Triticum turgidum ssp. *durum*	Hard wheat	grain kernel
–	egg	
–	salt	
–	water	

Impurities

–	insect

Description

The dough contains small fragments of bran and is flattened and cut into small, elongated units. The bran gives the pasta a somewhat crumbly texture. Both thickness and width differ to some extent. Scorch marks are present and are indicative of dry-heating. Weevils are present. This may indicate that the pasta was not heated, and that the infestation with eggs was present in the meal that was used. Alternatively, the infestation could have occurred sometime after processing, which took place 16 months earlier.

FIGURE 194

Pasta

Obtained	Sanandaj (Kurdistan province, Iran)
Origin	–
Date	April 2017
Collector(s)	R. T. J. Cappers & M. Kaffashi
Context	Bazaar
GIA-accession	40758
Length/Diameter	–
Thickness	–
Weight	–
Quantity	–
Photographer	D. Fennema

Composition

Triticum	Wheat	grain kernel
–	salt	
–	water	

Description

This pasta is home-made. The outer surface is covered with some flour to prevent it from sticking to the substrate while it is being flattened. Some pieces are brown to brownish black, indicating that the dough is dry-heated.

FIGURE 195

Pasta

Obtained	Tabriz (East Azerbaijan province, Iran)
Origin	Varzeghan (East Azerbaijan province, Iran)
Date	July 2017
Collector(s)	R. T. J. Cappers & M. Kaffashi
Context	Bazaar
GIA-accession	40841
Length/Diameter	–
Thickness	–
Weight	–
Quantity	–
Photographer	D. Fennema

Composition

Triticum	Wheat	grain kernel
–	salt	
–	water	

Description

This pasta is home-made. Several flattened pieces of dough are stacked before the dough is cut into elongated pieces. The somewhat spongy surface indicates that proving has taken place. The presence of scorch marks indicates that the pasta has been dry-heated.

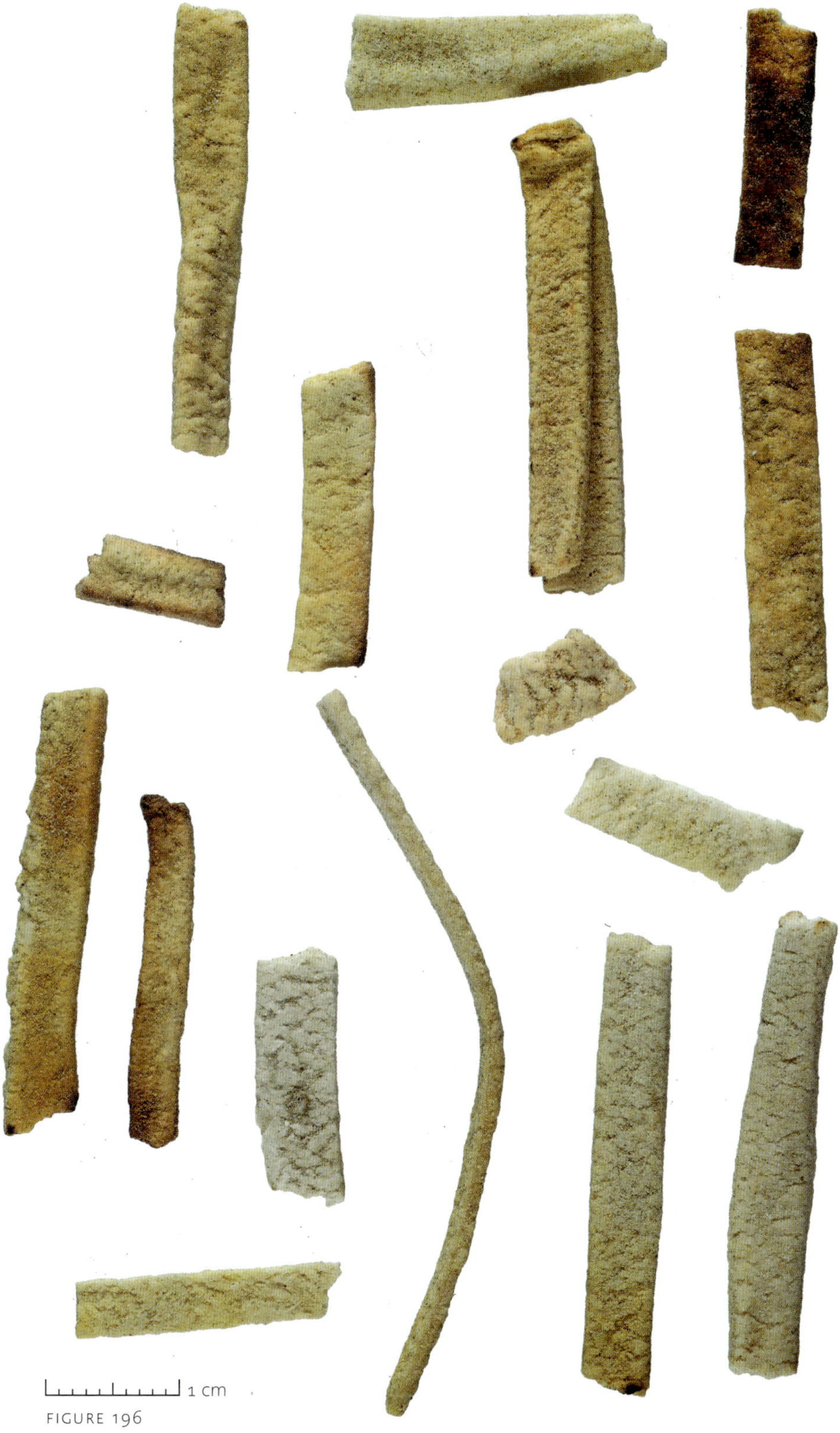

FIGURE 196

Pasta

Obtained	Düzce (Düzce province, Turkey)
Origin	Aydınıpınar (Düzce province, Turkey)
Date	February 2017
Collector(s)	R. T. J. Cappers, H. Balcı & L. Kutlu
Context	Farmers market
GIA-accession	40293
Length/Diameter	–
Thickness	–
Weight	–
Quantity	–
Photographer	D. Fennema

Composition

Triticum turgidum ssp. *durum*	Hard wheat	grain kernel
–	egg	
–	milk	
–	salt	
–	water	

Description

This pasta is home-made. The pasta is designated as '*erişte macaroni*' ('stripped pasta'), referring to the final shape. In addition to water, a small amount of cooked milk is used. After the dough has been shaped into elongated units, the pasta is boiled for 15 minutes and then sun-dried. The pasta can be stored for 2 months. The boiling will kill any eggs of insects and mites are present in the meal and thus reduce the chance of the pasta spoiling.

FIGURE 197

Pasta

Obtained	Hendek (Sakarya province, Turkey)
Origin	Gerede (Bolu province, Turkey)
Date	February 2017
Collector(s)	R. T. J. Cappers & H. Balcı
Context	Shop
GIA-accession	40302
Length/Diameter	–
Thickness	–
Weight	–
Quantity	–
Photographer	D. Fennema

Composition

Solanum lycopersicum	Tomato	fruit
Triticum turgidum ssp. *durum*	Hard wheat	grain kernel
–	egg	
–	salt	
–	water	

Description

The dough is flattened and cut into small, elongated units. Both thickness and width differ to some extent. Some broader units are present as well. The use of tomato has resulted in the orange colour. Scorch marks are present. The shelf life is 2 years according to the packaging.

FIGURE 198

Pasta

Obtained	Istanbul (Istanbul province, Turkey)
Origin	Kastamonu (Kastamonu province, Turkey)
Date	February 2017
Collector(s)	R. T. J. Cappers, M. Ozyarkent & H. Balcı
Context	Farmers market
GIA-accession	40324
Length/Diameter	–
Thickness	–
Weight	–
Quantity	–
Photographer	D. Fennema

Composition

Triticum turgidum ssp. *durum*	Hard wheat	grain kernel
–	salt	
–	water	

Description

This pasta is home-made. Scorch marks are absent, indicating that the flattened dough was not dry-heated prior to shaping. Concentrations of flour are present at the surface. This pasta is designated as '*üçgen eriste*' ('triangle pasta'). The triangles of this pasta are rather large. A similar kind of pasta was obtained in a shop in Kastamonu (Kastamonu province).

FIGURE 199

Pasta

Obtained	Düzce (Düzce province, Turkey)
Origin	Konuralp (Düzce province, Turkey)
Date	February 2017
Collector(s)	R. T. J. Cappers, H. Balcı & L. Kutlu
Context	Farmers market
GIA-accession	40292
Length/Diameter	–
Thickness	–
Weight	–
Quantity	–
Photographer	D. Fennema

Composition

Triticum turgidum ssp. *durum*	Hard wheat	grain kernel
–	egg	
–	milk	
–	salt	
–	water	

Description

This pasta is home-made. In addition to water, also a small amount of cooked milk is used. After shaping the dough is cut into triangular units. The pasta is not boiled and sun-dried. The pasta can be stored for 2 months.

FIGURE 200

Pasta

Obtained	Kalamafka (Crete region, Greece)
Origin	–
Date	July 2016
Collector(s)	R. T. J. Cappers
Context	Private house
GIA-accession	39892
Length/Diameter	–
Thickness	–
Weight	–
Quantity	–
Photographer	D. Fennema

Composition

Triticum turgidum ssp. *durum*	Hard wheat	grain kernel
–	salt	
–	water	

Description

This pasta is home-made. The dough is shaped into a long cylinder with the hands and then this cylinder is cut into small pieces. Each piece is first flattened and then rolled up with the fingertips. This pasta has been boiled for a period of time that is less than the full cooking time.

FIGURE 201

Pasta

Obtained	Kritsa (Crete region, Greece)
Origin	–
Date	July 2016
Collector(s)	R. T. J. Cappers
Context	Private house
GIA-accession	39891
Length/Diameter	–
Thickness	–
Weight	–
Quantity	–
Photographer	D. Fennema

Composition

Triticum turgidum ssp. *durum*	Hard wheat	grain kernel
–	salt	
–	water	

Impurities

–	insect

Description

This pasta is home-made. The dough is shaped into a long cylinder with the hands and then this cylinder is broken into small pieces. Each piece is first flattened and then rolled up around a metal pin. This pasta has been collected after this stage of the processing and has not yet been boiled. The presence of weevils after some time during storage indicates that insect eggs must have been present in the meal.

FIGURE 202

Pasta

Obtained	Berlin (GFS of Berlin, Germany)
Origin	–
Date	December 2014
Collector(s)	R. Neef
Context	Shop
GIA-accession	35116
Length/Diameter	–
Thickness	–
Weight	–
Quantity	–
Photographer	R. T. J. Cappers

Composition

Triticum turgidum ssp. durum	Hard wheat	grain kernel
–	salt	
–	water	

Description

A spiral-shaped pasta industrially made based on traditional processing. Until the recent past, poor people were allowed to gather spikes from the field after harvesting. For example, in the scene dealing with harvesting depicted on the wall of Paheri's mastaba (Egypt, New Kingdom), women complain about the accuracy of harvesting, which has resulted in insufficient spikes to gather. And in Belgium, the gleaning was established by law, describing the conditions of the time period that gleaning was allowed. When the stubble field was burnt, in order to reduce the viability of the diaspores of the arable weeds and to improve the mineralization, any spikes that were still present in the stubble field became roasted. It is from this kind of roasted grain that pasta was made. Modern pastas are usually not heated before storage or only partly boiled. The application of dry-heating makes this pasta rather exceptional. This pasta has a brownish colour and smoky flavour, and the dry-heating has the advantage that insects and eggs are killed and its shelf life is improved. According to the packaging, the dough is made from only meal and water.

1 cm

FIGURE 203

Pasta

Obtained	Groningen (province of Groningen, the Netherlands)
Origin	Sicily (Italy)
Date	July 2013
Collector(s)	R. T. J. Cappers & F. B. J. Heinrich
Context	Shop
GIA-accession	30771
Length/Diameter	–
Thickness	–
Weight	–
Quantity	–
Photographer	D. Fennema

Composition

Triticum turgidum ssp. *durum*	Hard wheat	grain kernel
–	salt	
–	water	

Description

This pasta is designated as '*macaroni*'. It is made from meal, which gives the pasta its typical brown colour. The shelf life according to the packaging is 2 years.

Nutritional value (100 g)

carbohydrates	70 g
dietary fibre	8 g
lipids	2 g
proteins	11.5 g
salt	4 mg

FIGURE 204

Canaan FAIR TRADE

PALESTINIAN FAIR TRADE

HAND ROLLED • SUN-DRIED

Canaan FAIR TRADE

ORGANIC MAFTOUL

PALESTINIAN COUSCOUS

WHOLE GRAIN • HAND ROLLED • SUN-DRIED

CANAAN FAIR TRADE DELICAIES
PREPARED USING ANCIENT METHODS
BY WOMEN-OWNED COOPERATIVES
IN PALESTINE

fair for life
fair trade certified by IMO

PS-BIO-143
Palestinian Agriculture

℮250g

13 Couscous

Only couscous in the strict sense is presented here: a dry intermediate food consisting of granules of a uniform size. It is made by mixing coarse and fine granulated grain with water. Salt can be added to improve the taste. In its basic form, pasta is made without heating.

The production of couscous can be considered complementary to that of bulgur. In both cereal foods, the producer of the food aims to obtaining a uniform size of granules that makes it possible to achieve doneness of all of the particles using the same amount of cooking time. Whereas in bulgur a uniform granule size is obtained by separating the mixture into different sizes of granules, in couscous a uniform granule size is obtained by merging the different-sized small granules into larger conglomerates with the aid of water. This merging of the different granules is done by stirring and is labour-intensive. In addition, the production and heating of couscous require special equipment. The producer should have access to a sieve with a rather small mesh size that is used for sieving the couscous granules into a standardized size fraction. The person doing the heating during food preparation, which is done by steaming, should have access to two pots: a lower one for boiling the water and an upper one with a perforated bottom for holding the couscous granules. These requirements suggest that production of couscous most probably did not take place on a large scale and only became possible once pottery had become available.

Couscous

Obtained	Groningen (province of Groningen, the Netherlands)
Origin	–
Date	July 2014
Collector(s)	F. Fantone
Context	Shop
GIA-accession	34832
Length/Diameter	–
Thickness	–
Weight	–
Quantity	–
Photographer	D. Fennema

Composition

Hordeum vulgare	Hulled Barley	grain kernels
–	water	

Description

The small-sized couscous is made from medium-ground grain kernels. The florets have been pearled prior to grinding. Small fragments of the bran, sometimes covered by the chaff (usually from the hilum groove, thus the palea) are present.

FIGURE 205

Couscous

Obtained	Utrecht (province of Utrecht, the Netherlands)
Origin	Palestine
Date	February 2017
Collector(s)	R. T. J. Cappers
Context	Shop
GIA-accession	40271
Length/Diameter	–
Thickness	–
Weight	–
Quantity	–
Photographer	D. Fennema

Composition

Triticum	Wheat	grain kernel
–	water	

Description

Couscous consisting of relatively large granules. The packaging mentions that bulgur has been used, which indicates that the grain was par-boiled prior to grinding. The granules are hand-rolled, likely by women, and sun-dried. The shape of the granules is rather variable and varies from almost spherical to somewhat elongated and irregular. The grain kernels are not polished. The shelf life indicated is 2 years.

FIGURE 206

Couscous

Obtained	Rabat (Rabat-Salé-Kénitra region, Morocco)
Origin	–
Date	March 2016
Collector(s)	R. T. J. Cappers
Context	Bazaar
GIA-accession	39840
Length/Diameter	–
Thickness	–
Weight	–
Quantity	–
Photographer	D. Fennema

Composition

Triticum turgidum ssp. *durum*	Hard wheat	grain kernel
–	water	

Description

Small-sized couscous made from finely ground, polished grain kernels. Bran fragments are absent.

FIGURE 207

Couscous

Obtained	Utrecht (province of Utrecht, the Netherlands)
Origin	Morocco
Date	February 2017
Collector(s)	R.T.J. Cappers
Context	Shop
GIA-accession	40264
Length/Diameter	–
Thickness	–
Weight	–
Quantity	–
Photographer	D. Fennema

Composition

Triticum turgidum ssp. *durum*	Hard wheat	grain kernel
–	water	

Description

Medium-sized couscous made from polished grain kernels. Only minute fragments of the bran are still present. Each granule consists of an agglomeration of smaller granules. The shelf life according to the packaging is 2 years.

FIGURE 208

Couscous

Obtained	Rabat (Rabat-Salé-Kénitra region, Morocco)
Origin	–
Date	March 2016
Collector(s)	R. T. J. Cappers
Context	Bazaar
GIA-accession	39833
Length/Diameter	–
Thickness	–
Weight	–
Quantity	–
Photographer	D. Fennema

Composition

Triticum turgidum ssp. *durum*	Hard wheat	grain kernel
–	water	

Description

Small-sized couscous made from finely ground grain kernels. Each composite unit consists of only a few granules. The grain kernels are polished before grinding, and almost no bran fragments are present.

FIGURE 209

Couscous

Obtained	Amsterdam (province of Noord Holland, the Netherlands)
Origin	Morocco
Date	September 2017
Collector(s)	R.T.J. Cappers
Context	Shop
GIA-accession	40983
Length/Diameter	–
Thickness	–
Weight	–
Quantity	–
Photographer	D. Fennema

Composition

Triticum turgidum ssp. *durum*	Hard wheat	grain kernel
–	water	

Impurities

–	insect	

Description

The couscous is made from polished and finely ground grain kernels. Fragments of the bran are present but rare. A live moth was present in the sample 5 months after the couscous was bought.

FIGURE 210

Couscous

Obtained	Amsterdam (province of Noord Holland, the Netherlands)
Origin	Aix en Provence (Provence-Alpes-Côte d'Azur, France)
Date	January 2016
Collector(s)	R.T.J. Cappers
Context	Shop
GIA-accession	39741
Length/Diameter	–
Thickness	–
Weight	–
Quantity	–
Photographer	D. Fennema

Composition

Triticum turgidum ssp. *durum*	Hard wheat	grain kernel
–	water	

Description

Couscous made from polished grain kernels. Bran fragments are not present. The packaging states that the contents is '*semolina couscous*', that the cooking time is 4 minutes, and that the shelf life is 2 years. The term '*semolina*' is used for finely ground grain.

Nutritional value (100 g)

carbohydrates	72 g
dietary fibre	4 g
lipids	1 g
proteins	13 g

FIGURE 211

Couscous

Obtained	Groningen (province of Groningen, the Netherlands)
Origin	Italy
Date	April 2014
Collector(s)	R. T. J. Cappers & F. Fantone
Context	Shop
GIA-accession	34704
Length/Diameter	–
Thickness	–
Weight	–
Quantity	–
Photographer	D. Fennema

Composition

Triticum turgidum ssp. *durum*	Hard wheat	grain kernel
–	water	

Description

This couscous consists of small, composite granules that have been par-boiled by steaming. The couscous is sold as '*volkoren couscous*'. '*Volkoren*' is the Dutch word for 'whole wheat', indicating that the couscous is made from unpolished grain kernels.

Nutritional value (100 g)

carbohydrates	65.7 g
dietary fibre	8.7g
lipids	2.7 g
proteins	13.8 g

FIGURE 212

Couscous

Obtained	Leeuwarden (province of Friesland, the Netherlands)
Origin	Ankara (Ankara province, Turkey)
Date	March 2016
Collector(s)	M. van Rijn
Context	Shop
GIA-accession	39945
Length/Diameter	–
Thickness	–
Weight	–
Quantity	–
Photographer	D. Fennema

Composition

Triticum turgidum ssp. *durum*	Hard wheat	grain kernel
–	water	

Description

Couscous pellets made from finely ground grain kernels (semolina). The semolina is pressed into a cylinder shape that is fragmented into small units with convex cross sections. The name given on the packaging is '*Ankara kuskus*'. The cooking time is 12–14 minutes. According to the packaging, the shelf life is 2 years.

Nutritional value (100 g)

carbohydrates	73.3 g
dietary fibre	2.9 g
lipids	1.7 g
proteins	10.7 g
salt	4.0 mg

FIGURE 213

Couscous

Obtained	Groningen (province of Groningen, the Netherlands)
Origin	–
Date	March 2005
Collector(s)	F. Fantone
Context	Shop
GIA-accession	39577
Length/Diameter	–
Thickness	–
Weight	–
Quantity	–
Photographer	D. Fennema

Composition

Triticum turgidum ssp. durum	Hard wheat	grain kernel
–	water	

Description

Mechanically made couscous. The pulverized grain kernels have been pressed into pill-shaped granules. The granules differ in both diameter and thickness.

FIGURE 214

کشک عشایری
کم نمک برای
خوردن و غذا
۱۲۴ - ۴۱۰

14 Kashk

Kashk is a dried, non-rennet cheese. Rennet is an extract of the stomach of a ruminant mammal that can be used to solidify the proteins in milk in order to make cheese. Non-rennet cheese is made without adding this extract; instead, the solidification of the proteins is solely the result of a lowering of the acidity due to lactic acid fermentation.

It is assumed that kashk was first produced once domestic mammals had started producing a surplus of milk. The production of kashk need not involve heating and requires only simple equipment, made of organic material. In the past, the production of kashk was most probably part of the second stage in milk processing. In the first stage, fat is extracted from the milk for making butter. In the second stage, the low-fat milk that remains (either skimmed milk or buttermilk) can be used for the production of kashk or for mixing with grain to produce tarhana. Some of the modern samples of kashk in the collection at GIA are made from sour milk (yoghurt), and thus retain most of the nutrients that were present in the milk originally.

The modern kashk samples differ in composition, shape, and size. The composition is determined by the kind of animal (sheep, goat, cow, or water buffalo) that produced the milk and the manner of processing of that milk. Unfortunately, the composition of the kashk samples could not always be documented at the time when they were being collected. In those cases where it could not be documented, the composition will have to be established through further analysis. In most cases, the kashk was divided and shaped by hand. Occasionally, a thread was used for cutting it into pieces.

Kashk

Obtained	Kandovan (East Azerbaijan province, Iran)
Origin	–
Date	June 2017
Collector(s)	R.T.J. Cappers & M. Kaffashi
Context	Shop
GIA-accession	40857
Length/Diameter	(5.2–) 5.5 (–6.0) cm
Thickness	(2.3–) 2.6 (–3.0) cm
Weight	(14.3–) 17.8 (–24.9) g
Quantity	4
Photographer	D. Fennema

Composition

–	sheep's milk

Description
The kashk was shaped in one hand, showing the imprints of the encircling fingers.

FIGURE 215

Kashk

Obtained	Tabriz (East Azerbaijan province, Iran)
Origin	Mahabad (West Azerbaijan province, Iran)
Date	July 2017
Collector(s)	R. T. J. Cappers & M. Kaffashi
Context	Bazaar
GIA-accession	40850
Length/Diameter	(5.0–) 5.6 (–6.4) cm
Thickness	(1.8–) 2.1 (–2.4) cm
Weight	(10.3–) 12.4 (–14.4) g
Quantity	4
Photographer	D. Fennema

Composition

–	cow's milk

Description
The kashk was shaped in one hand, showing the encircling imprints of the fingers. Some additional decoration is added in the form of lines and cross-hatching.

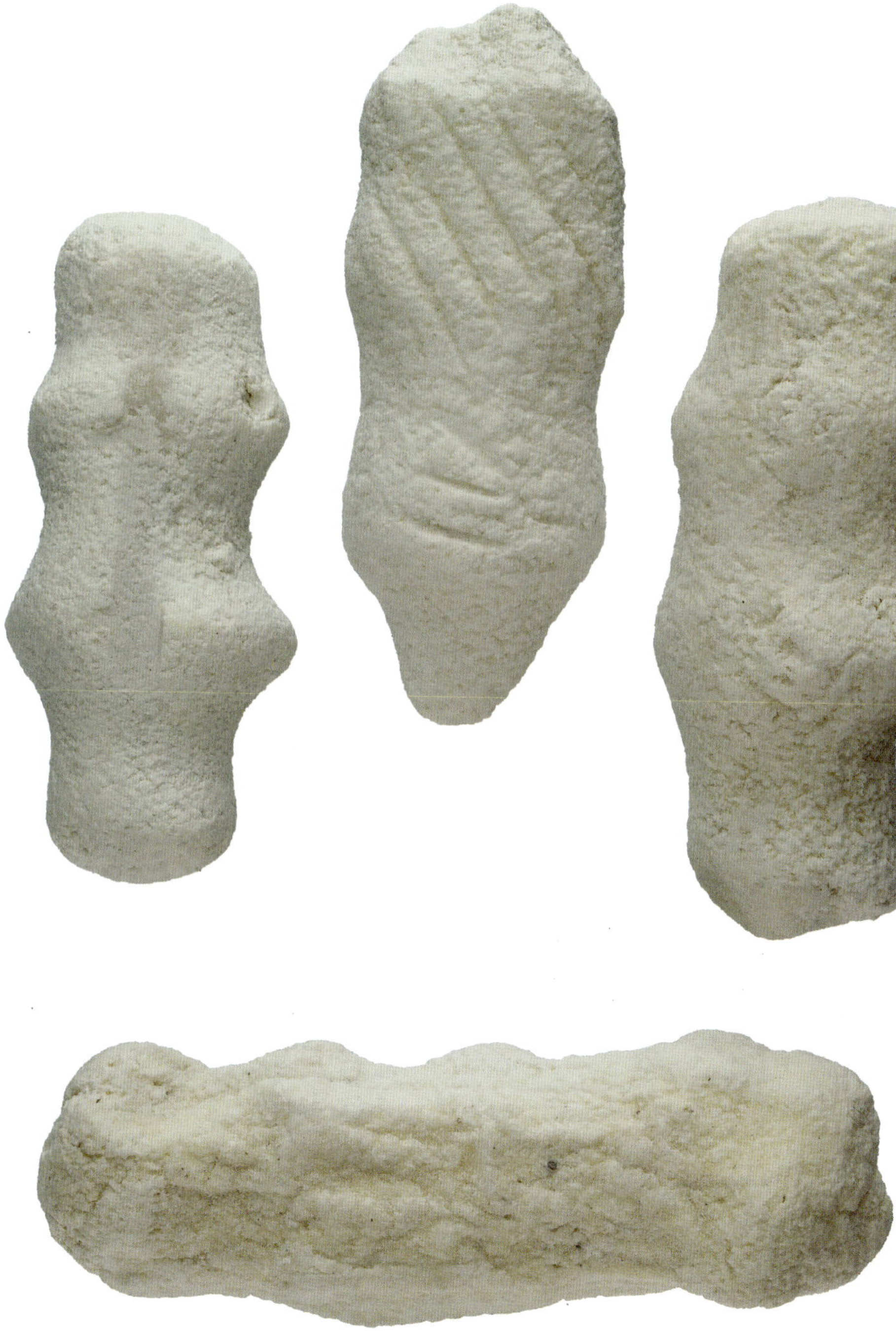

FIGURE 216

Kashk

Obtained	Tabriz (East Azerbaijan province, Iran)
Origin	–
Date	July 2017
Collector(s)	R.T.J. Cappers & M. Kaffashi
Context	Bazaar
GIA-accession	40859
Length/Diameter	(4.6–) 5.5 (–6.6) cm
Thickness	(1.9–) 2.0 (–2.2) cm
Weight	(16.7–) 20.1 (–22.9) g
Quantity	6
Photographer	D. Fennema

Composition

–	sheep's milk

Description

The kashk was shaped in one hand into elongated units. Imprints of the fingers and palm are visible. The units have a clearly noticeable flat side, where they have been pushed into the drying surface. The quality is considered inferior due to both its taste and its smell.

FIGURE 217

Kashk

Obtained	Tabriz (East Azerbaijan province, Iran)
Origin	–
Date	July 2017
Collector(s)	R. T. J. Cappers & M. Kaffashi
Context	Bazaar
GIA-accession	40870
Length/Diameter	(4.9–) 5.0 (–5.3) cm
Thickness	(1.4–) 1.5 (–1.6) cm
Weight	(9.3–) 10.7 (–12.2) g
Quantity	3
Photographer	D. Fennema

Composition

–	cow's milk

Description
The kashk was shaped in one hand and shows the imprints of the fingers on one face and the imprint of the palm on the other face.

FIGURE 218

Kashk

Obtained	Urmia (East Azerbaijan province, Iran)
Origin	Miyaneh (East Azerbaijan province, Iran)
Date	July 2017
Collector(s)	J. Abdeli
Context	Shop
GIA-accession	41007
Length/Diameter	6.2 cm
Thickness	2.7 cm
Weight	31.3 g
Quantity	1
Photographer	D. Fennema

Composition

–	sheep's milk

Description

The kashk was shaped in one hand into an elongated unit. No finger impressions are visible. The bottom face is flattened against the drying surface. The kashk was made by a Mrs. Ghorbani. She mentioned that the kashk has to be grated into a powdery substance. This ground kashk is then soaked in water. It is also heated to prevent brucellosis.

FIGURE 219

Kashk

Obtained	Urmia (East Azerbaijan province, Iran)
Origin	West Azerbaijan province, Iran
Date	October 2017
Collector(s)	J. Abdeli
Context	Shop
GIA-accession	41009
Length/Diameter	(6.5–) 7.1 (–7.5) cm
Thickness	(2.2–) 2.7 (–3.1) cm
Weight	(27.5–) 29.1 (–31.4) g
Quantity	4
Photographer	D. Fennema

Composition

–	cow's milk

Description

The kashk was shaped in one hand into an elongated unit. A broad, flat side is created by pushing the units into the drying surface. The quality is considered inferior due to both its taste and its smell. It was made some 6 months before it was acquired.

FIGURE 220

Kashk

Obtained	Bulber (Kurdistan province, Iran)
Origin	–
Date	April 2017
Collector(s)	R.T.J. Cappers, M. Kaffashi & P. Hesami
Context	Tourist shop
GIA-accession	40767
Length/Diameter	(5.1–) 5.6 (–6.0) cm
Thickness	(1.1–) 1.3 (–1.7) cm
Weight	(5.8–) 7.4 (–9.1) g
Quantity	10
Photographer	D. Fennema

Composition

–	milk

Description
The kashk was shaped in one hand, and the imprints of the fingers are slightly visible. The imprint of the palm has been replaced by that of the flat work surface. This kashk was obtained in a shop selling tourist items that also offers some traditional foods for sale.

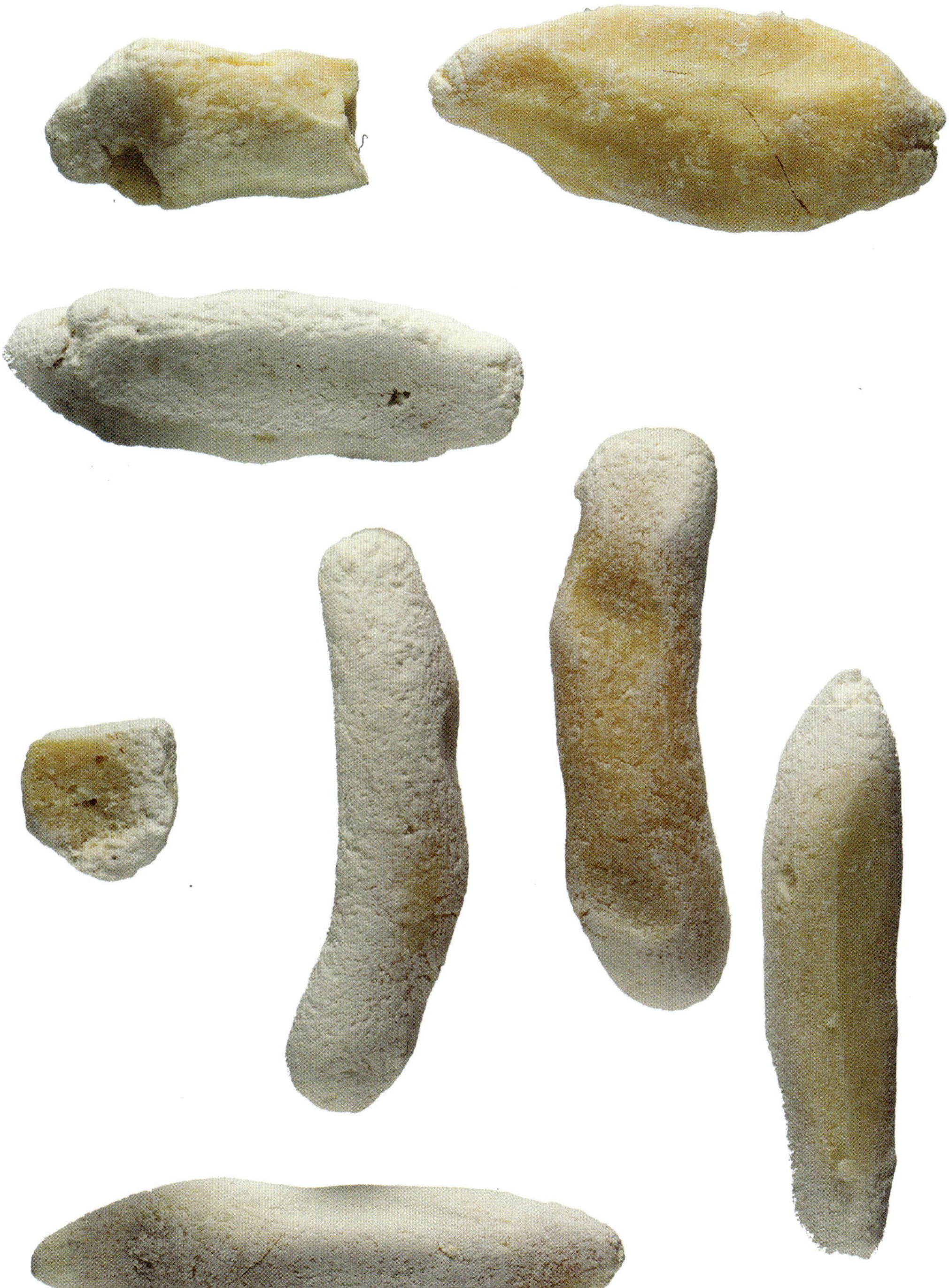

FIGURE 221

Kashk

Obtained	Hamedan (Hamedan province, Iran)
Origin	Marivan (Kurdistan province, Iran)
Date	July 2017
Collector(s)	R.T.J. Cappers & M. Kaffashi
Context	Bazaar
GIA-accession	40858
Length/Diameter	(4.4–) 4.7 (–5.0) cm
Thickness	(1.3–) 1.4 (–1.5) cm
Weight	(6.3–) 7.1 (–7.4) g
Quantity	3
Photographer	D. Fennema

Composition

–	sheep's milk

Description

The kashk was shaped in one hand, but the impression of fingers and palm are almost invisible. The texture is somewhat crumbly. The bottom face, facing the drying surface, is flattened, and one piece shows a clear imprint of cross-hatching.

FIGURE 222

Kashk

Obtained	Hamedan (Hamedan province, Iran)
Origin	–
Date	July 2017
Collector(s)	R. T. J. Cappers & M. Kaffashi
Context	Bazaar
GIA-accession	41001
Length/Diameter	4.4 cm
Thickness	1.6 cm
Weight	8.0 g
Quantity	1
Photographer	D. Fennema

Composition

–	sheep's milk

Description
The kashk was rolled with both hands into a cylindrical unit with rounded ends.

FIGURE 223

Kashk

Obtained	Shiraz (Fars province, Iran)
Origin	–
Date	July 2017
Collector(s)	R. T. J. Cappers & M. Kaffashi
Context	Shop
GIA-accession	40836
Length/Diameter	(3.0–) 3.3 (–3.5) cm
Thickness	(0.9–) 1.3 (–1.7) cm
Weight	(7.7–) 9.4 (–11.7) g
Quantity	4
Photographer	D. Fennema

Composition

–	sheep's milk

Description
The kashk is first shaped into a thick roll, with a diameter of c. 3.3 cm. Next, this role is dissected into small parts, each with a width of c. 1.3 cm. The units are dried flat face down, and imprints of the drying surface can be seen.

FIGURE 224

Kashk

Obtained	Kandovan (East Azerbaijan province, Iran)
Origin	–
Date	June 2017
Collector(s)	R. T. J. Cappers & M. Kaffashi
Context	Shop
GIA-accession	40854
Length/Diameter	(3.4–) 4.1 (–4.5) cm
Thickness	(2.2–) 2.5 (–3.0) cm
Weight	(19.6–) 22.5 (–29.6) g
Quantity	6
Photographer	D. Fennema

Composition

–	sheep's milk

Description

The kashk was shaped into relatively large, globular units that have become flattened on their bases.

FIGURE 225

Kashk

Obtained	Kandovan (East Azerbaijan province, Iran)
Origin	–
Date	June 2017
Collector(s)	R. T. J. Cappers & M. Kaffashi
Context	Shop
GIA-accession	40853
Length/Diameter	4.6 cm
Thickness	3.4 cm
Weight	35.8 g
Quantity	1
Photographer	D. Fennema

Composition

–	sheep's milk

Description
This kashk consists of a relatively large, globular unit that was shaped with both hands. The bottom shows a deep imprint that seems to be related to the drying surface.

FIGURE 226

Kashk

Obtained	Shiraz (Fars province, Iran)
Origin	–
Date	July 2017
Collector(s)	R. T. J. Cappers & M. Kaffashi
Context	–
GIA-accession	40849
Length/Diameter	(2.1–) 2.3 (–2.4) cm
Thickness	–
Weight	(4.8–) 5.4 (–5.8) g
Quantity	4
Photographer	D. Fennema

Composition

–	sheep's milk

Description
The kashk was shaped into globular units. They have been placed close to each other during the drying process, resulting in some flat edges.

FIGURE 227

Kashk

Obtained	Tabriz (East Azerbaijan province, Iran)
Origin	–
Date	July 2017
Collector(s)	R.T.J. Cappers & M. Kaffashi
Context	Bazaar
GIA-accession	40889
Length/Diameter	(3.0–) 3.3 (–3.5) cm
Thickness	(2.7–) 2.8 (–2.9) cm
Weight	(15.4–) 17.2 (–18.6) g
Quantity	4
Photographer	D. Fennema

Composition

–	sheep's milk

Description
The kashk was shaped into medium-sized, globular units using both hands. Two different kinds of imprints are present. One kind consists of small, parallel striations, most probably made with a fork. The other kind consists of a diamond shape, probably made with the handle of a fork. Such decorations may have been added to recall the imprints that were formerly produced by drying kashk on reed culms.

FIGURE 228

Kashk

Obtained	Urmia (East Azerbaijan province, Iran)
Origin	Golparchin (West Azerbaijan province, Iran)
Date	October 2017
Collector(s)	J. Abdeli
Context	Shop
GIA-accession	41008
Length/Diameter	(1.8–) 2.0 (–2.2) cm
Thickness	(1.4–) 1.7 (–2.1) cm
Weight	(3.6–) 4.1 (–4.9) g
Quantity	7
Photographer	D. Fennema

Composition

–	milk

Description

The kashk was shaped by rolling it between both hands into globular units. This kashk was made in September 2017 and photographed shortly thereafter.

FIGURE 229

Kashk

Obtained	Tehran (Tehran province, Iran)
Origin	Malayer (Hamedan province, Iran)
Date	January 2018
Collector(s)	S. Kamjan
Context	Shop
GIA-accession	41120
Length/Diameter	(2.1–) 2.2 (–2.4) cm
Thickness	(1.7–) 1.9 (–2.0) cm
Weight	(5.2–) 5.7 (–6.1) g
Quantity	4
Photographer	D. Fennema

Composition

Mentha	Mint	leaf
–	sheep's milk	

Description

The kashk was rolled into small units with both hands. It is flavoured with some dried mint.

FIGURE 230

Kashk

Obtained	Shiraz (Fars province, Iran)
Origin	–
Date	July 2017
Collector(s)	R.T.J. Cappers & M. Kaffashi
Context	Bazaar
GIA-accession	40845
Length/Diameter	(3.9–) 4.5 (–5.1) cm
Thickness	(0.9–) 1.1 (–1.2) cm
Weight	(7.1–) 7.7 (–8.7) g
Quantity	4
Photographer	D. Fennema

Composition

–	cow's milk

Description
The kashk was shaped into triangular units using both hands and a thread. First, the kashk is shaped into a large, elongated mass with a gable-shaped top. With a thread, vertical incisions are made at intervals of 1 cm. Then the top is dissected by pulling the thread horizontally. This is done in such a way that triangular units of similar size are obtained. When these have been removed, the large mass is moulded again to obtain a new top that can be dissected in a similar way. When small, hard particles are present, the thread causes a characteristic line on the cut face. In the product shown in the photograph, both flat sides and one small side were cut with the thread, whereas the two small sides were moulded by the hands. The kashk seems to be spiced.

FIGURE 231

Kashk

Obtained	Shiraz (Fars province, Iran)
Origin	Abadeh – Eqlid region, Iran
Date	July 2017
Collector(s)	R. T. J. Cappers & M. Kaffashi
Context	Bazaar
GIA-accession	40848
Length/Diameter	(4.1–) 4.9 (–5.8) cm
Thickness	(1.4–) 1.7 (–2.0) cm
Weight	(8.3–) 12.6 (–14.9) g
Quantity	5
Photographer	D. Fennema

Composition

–	sheep's milk

Description

The triangle-shaped kashk is originally made by moulding the solidified kashk into a large, elongated, triangular mass. The top is cut with a thread lengthwise, so that each side is approximately the same dimensions. Next, small pieces are obtained by cutting the kashk with the thread vertically, from one side to the other. The combination of hand-shaping two sides and cutting the third with a thread can be recognized in specimens if their shape has not changed too much due to the drying: two sides will be rather irregular and somewhat concave, and the third side will be more straight. Shaping the quantity of the fresh non-rennet cheese in this manner takes less time than making balls of it by rolling between both hands.

FIGURE 232

Kashk

Obtained	Shiraz (Fars province, Iran)
Origin	Eqlid and Abadeh (Fars province, Iran)
Date	July 2017
Collector(s)	R. T. J. Cappers & M. Kaffashi
Context	Bazaar
GIA-accession	40837
Length/Diameter	4.9–6.1 cm
Thickness	1.1–1.2 cm
Weight	8.3–10.3 g
Quantity	2
Photographer	D. Fennema

Composition

Origanum	Oregano	leaf
–	sheep's milk	

Description

The kashk is spiced with mountain oregano. The characteristic light yellow colour of the oregano is present in freshly made kashk and vanishes gradually over time.

FIGURE 233

Kashk

Obtained	Tehran (Tehran province, Iran)
Origin	Malayer (Hamedan province, Iran)
Date	January 2018
Collector(s)	S. Kamjan
Context	–
GIA-accession	41118
Length/Diameter	(2.6–) 2.8 (–2.9) cm
Thickness	(1.5–) 1.7 (–1.9) cm
Weight	(5.1–) 5.5 (–5.8) g
Quantity	4
Photographer	D. Fennema

Composition

–	buffalo milk

Description

The kashk was shaped by using a piping bag. The kashk must have been rather solid when it was put in the bag, judging from its final shape.

FIGURE 234

Kashk

Obtained	Kandovan (East Azerbaijan province, Iran)
Origin	Shop
Date	June 2017
Collector(s)	R. T. J. Cappers & M. Kaffashi
Context	Shop
GIA-accession	40860
Length/Diameter	(2.2–) 2.5 (–2.8) cm
Thickness	(1.0–) 1.3 (–1.6) cm
Weight	(2.2–) 2.9 (– 3.6) g
Quantity	3
Photographer	D. Fennema

Composition

–	milk

Description

The kashk consists of rather small units. The kashk was shaped by pressing it with the fingertips against the work surface. Only one unit bears the imprint of a cross. It is not clear whether this is from the structure of the work surface or added with a tool.

FIGURE 235

Kashk

Obtained	Tabriz (East Azerbaijan province, Iran)
Origin	–
Date	July 2017
Collector(s)	R. T. J. Cappers & M. Kaffashi
Context	Bazaar
GIA-accession	40814
Length/Diameter	(3.2–) 3.8 (–4.4) cm
Thickness	(1.9–) 2.3 (–2.4) cm
Weight	(11.1–) 14.4 (–16.4) g
Quantity	6
Photographer	D. Fennema

Composition

–	sheep's milk

Description
The kashk was shaped with the fingertips of both hands into irregular-shaped units. The surface has the same colour as the inner part. It appears from the colour that this kashk was made with milk from which no fat had been extracted. A flat face is created by pushing the units against the drying surface.

FIGURE 236

Kashk

Obtained	Tikan (Isfahan province, Iran)
Origin	–
Date	March 2018
Collector(s)	M. Djamali
Context	Private house
GIA-accession	41150
Length/Diameter	(3.4–) 3.6 (–3.9) cm
Thickness	(1.3–) 1.6 (–1.7) cm
Weight	(8.3–) 8.8 (–9.7) g
Quantity	4
Photographer	D. Fennema

Composition

–	sheep's yoghurt

Description
This kashk is made by first boiling the yoghurt for a number of hours. In this way, most of the water is removed by evaporation. About 5–6 litres of milk is needed to make 1 kg of kashk. After the yoghurt has been boiled down, the resulting sticky paste is divided into small units and shaped by pressing it with the hands, which are used like a pastry bag. The shaped units are then sun-dried and care is taken to keep birds away by using humans or a scarecrow. The colour is somewhat dark due to the long heating time. This kashk was made in spring 2017. In Tikan, kashk is also made from goat's milk and cow's milk.

FIGURE 237

Kashk

Obtained	Tabriz (East Azerbaijan province, Iran)
Origin	Mashhad (Razavi Khorasan province, Iran)
Date	June 2017
Collector(s)	R. T. J. Cappers & M. Kaffashi
Context	Bazaar
GIA-accession	40957
Length/Diameter	0.9 cm
Thickness	–
Weight	–
Quantity	–
Photographer	D. Fennema

Composition

–	milk

Description

An industrially produced kashk made into long, cylindrical units.

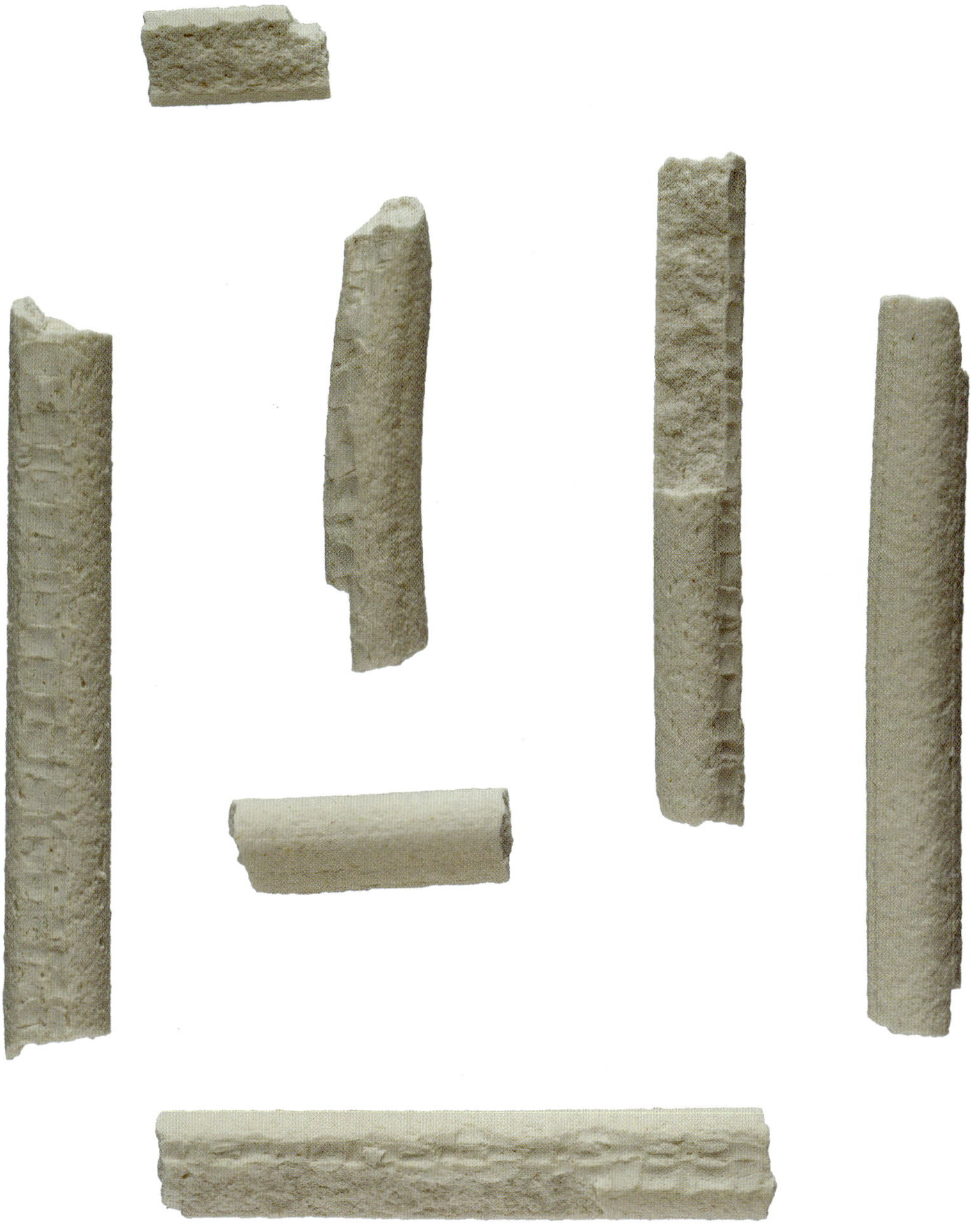

FIGURE 238

Kashk

Obtained	Amman (Amman governorate, Jordan)
Origin	Karak (Karak governorate, Jordan)
Date	September 2017
Collector(s)	R.T.J. Cappers
Context	Shop
GIA-accession	41027
Length/Diameter	6.3 cm
Thickness	6.2 cm
Weight	139.8 g
Quantity	1
Photographer	D. Fennema

Composition

–	sheep's (?) yoghurt
–	salt

Description

This unit of kashk has a large, conical shape, most probably involving Sheep's milk. The milk is made into yoghurt and this is mixed with salt. Next, the excess water is extracted, either by gravity or by pressing it for a period of 3–4 days. Then the kashk is moulded with both hands and allowed to finish drying for a period of 8–10 days. Its local name in Arabic is *'jameed'*.

FIGURE 239

Kashk

Obtained	Beverwijk (province of Noord Holland, the Netherlands)
Origin	Ghorband (Parwan province, Afghanistan)
Date	September 2017
Collector(s)	R.T.J. Cappers
Context	Shop
GIA-accession	41023
Length/Diameter	(4.2–) 4.8 (–5.6) cm
Thickness	(4.0–) 4.2 (–4.3) cm
Weight	(23.2–) 28.3 (–36.8) g
Quantity	5
Photographer	D. Fennema

Composition

–	milk

Description
The kashk was shaped into relatively large units that consist of two conical parts that are joined together at their broad, circular face. Although this shape might take some time to model, it is quite efficient in storage. The compaction is relatively high, and the empty space between the units is small. These units have been separated from two large bags that were offered for sale and most probably concern a mixture of leftovers of different origins. The imprints on one side are probably from the drying surface.

FIGURE 240

Kashk

Obtained	Beverwijk (province of Noord Holland, the Netherlands)
Origin	Ghorband (Parwan province, Afghanistan)
Date	September 2017
Collector(s)	R. T. J. Cappers
Context	Shop
GIA-accession	41027
Length/Diameter	3.9–4.4 cm
Thickness	3.1–3.6 cm
Weight	26.7–39.1 g
Quantity	2
Photographer	D. Fennema

Composition

–	milk

Description
This kashk was shaped into relatively large, globular units by rolling between both hands. Some incised lines are present on one face, but their orientation rules out that they originate from the drying surface.

FIGURE 241

15
Qara qurut

Qara qurut is solid or dry whey. Qara qurut is classified here as a unique traditional food. Despite its common name 'black cheese', it is not a cheese in the strict sense. Today, its production seems to be geographically limited, and the examples that are depicted here originate from Iran and Afghanistan. It is questionable to what extent this milk product was produced in the past. In comparison to butter and kashk, qara qurut is maybe not a good trade-off. Although the whey consists of valuable nutrients, including milk sugar, acids, and whey proteins, it takes a lot of fuel to boil down the whey to a solid product that can be dried and stored for a long time.

The first examples that are depicted in the atlas concern a special kind of dry milk product because they consist of one part kashk (non-rennet cheese, white colour) and one part qara qurut (whey, brown-black colour), which have been combined into a single unit after their individual processing. The succeeding examples that are depicted concern pure quantities of qara qurut. These are offered for sale as large blocks, from which the desired quantity can be extracted, or as pre-packaged, smaller units.

Kashk & qara qurut

Obtained	Shiraz (Fars province, Iran)
Origin	–
Date	July 2017
Collector(s)	R.T.J. Cappers & M. Kaffashi
Context	Bazaar
GIA-accession	40844
Length/Diameter	2.4–2.5 cm
Thickness	(1.6–) 1.7 (1.9) cm
Weight	(5.2–) 5.4 (–5.6) g
Quantity	3
Photographer	D. Fennema

Composition

–	kashk
–	qara qurut (whey)

Description

The globular units consist of a combination of kashk (white part) and qara qurut (dark part). The light brown part seems to be a mixture of kashk and qara qurut, and hence this product is considered a lower quality than a product in which pure qara qurut is used for the dark part.

FIGURE 242

Kashk & qara qurut

Obtained	Isfahan (Isfahan province, Iran)
Origin	Shahrekord (Chaharmahal and Bakhtiari province, Iran)
Date	April 2017
Collector(s)	R. T. J. Cappers & M. Kaffashi
Context	Bazaar
GIA-accession	40427
Length/Diameter	(2.2–) 2.3 (–2.4) cm
Thickness	(2.0–) 2.1 (–2.3) cm
Weight	(6.7–) 7.3 (–8.6) g
Quantity	6
Photographer	D. Fennema

Composition

–	kashk
–	qara qurut (whey)

Description

The globular units consist of a combination of kashk (white part) and qara qurut (dark part). The kashk exhibits small, light brown spots where it has come into contact with the qara qurut of adjacent units.

FIGURE 243

Kashk & qara qurut

Obtained	Isfahan (Isfahan province, Iran)
Origin	–
Date	April 2017
Collector(s)	R. T. J. Cappers & M. Kaffashi
Context	Bazaar
GIA-accession	40437
Length/Diameter	(2.4–) 2.6 (–3.5) cm
Thickness	2.1–2.2 cm
Weight	(7.7–) 8.3 (–9.4) g
Quantity	6
Photographer	D. Fennema

Composition

–	kashk
–	qara qurut (whey)

Description

The globular units consist of a combination of kashk (white part) and qara qurut (dark part).

FIGURE 244

Kashk & qara qurut

Obtained	Shiraz (Fars province, Iran)
Origin	–
Date	July 2017
Collector(s)	R. T. J. Cappers & M. Kaffashi
Context	–
GIA-accession	40800
Length/Diameter	(2.2–) 2.4 (–2.5) cm
Thickness	–
Weight	(5.5–) 6.8 (–7.9) g
Quantity	5
Photographer	D. Fennema

Composition

–	kashk
–	qara qurut (whey)

Description

The globular units consist of a combination of kashk (white part) and qara qurut (dark part).

FIGURE 245

Qara qurut

Obtained	Urmia (East Azerbaijan province, Iran)
Origin	West Azerbaijan Province, Iran
Date	October 2017
Collector(s)	J. Abdeli
Context	Shop
GIA-accession	41003
Length/Diameter	4.9–6.5 cm
Thickness	2.2–2.7 cm
Weight	14.4–33.5 g
Quantity	2
Photographer	D. Fennema

Composition

–	whey

Description
This qara qurut originates from a small village south of Urmia. It is slightly brown yet is designated as white qara qurut. Its quality is considered high because it is not extremely sour and thus can be eaten in rather large pieces without causing an unpleasant sensation in the mouth. This qara qurut was probably 1 year old when it was obtained.

FIGURE 246

Qara qurut

Obtained	Ardabil (Ardabil province, Iran)
Origin	–
Date	July 2017
Collector(s)	R. T. J. Cappers & M. Kaffashi
Context	Bazaar
GIA-accession	40831
Length/Diameter	3.5–3.8 cm
Thickness	2.0–2.9 cm
Weight	8.3–13.0 g
Quantity	2
Photographer	D. Fennema

Composition

–	whey

Description

These fragments were taken from a large, square block. Such blocks are currently made on an industrial scale. The colour is almost blackish and is predictive of a pronounced sour taste.

FIGURE 247

Qara qurut

Obtained	Iran
Origin	–
Date	2017
Collector(s)	S. Y. F. Daghati
Context	–
GIA-accession	41123
Length/Diameter	13 cm
Thickness	1.8 cm
Weight	220.5 g
Quantity	1
Photographer	D. Fennema

Composition

–	whey

Description
This elongated block of qara qurut was offered for sale pre-packed in plastic. The colour is dark brown and predictive of a strong taste.

FIGURE 248

Qara qurut

Obtained	Beverwijk (province of Noord Holland, the Netherlands)
Origin	Afghanistan
Date	September 2017
Collector(s)	R. T. J. Cappers
Context	Shop
GIA-accession	40992
Length/Diameter	3.2 cm
Thickness	1.2–13. cm
Weight	7.1–7.6 g
Quantity	2
Photographer	D. Fennema

Composition

–	whey

Description

These rod-shaped units of qara qurut were originally individually wrapped in plastic. The colour is dark brown and its taste is extremely sour. According to the shop owner, this qara qurut is especially eaten by women who want to lose weight.

FIGURE 249

Qara qurut

Obtained	Isfahan (Isfahan province, Iran)
Origin	Shahrekord (Chaharmahal and Bakhtiari province, Iran)
Date	April 2017
Collector(s)	R. T. J. Cappers & M. Kaffashi
Context	Bazaar
GIA-accession	40426
Length/Diameter	(3.0–) 3.4 (–3.7) cm
Thickness	(1.3–) 1.5 (–1.8) cm
Weight	(6.6–) 7.4 (–9.0) g
Quantity	8
Photographer	D. Fennema

Composition

–	whey

Description

The qara qurut has been shaped by using a pastry bag. The qara qurut must still have been quite liquid, judging from its final shape. The smell is rather distinctive, which may be the result of the use of certain spices or of sub-optimal storage. A similar-shaped qara qurut, originating from Semiron (Isfahan province), was also obtained in the bazaar of Isfahan.

FIGURE 250

Qara qurut

Obtained	Beverwijk (province of Noord Holland, the Netherlands)
Origin	Iran
Date	September 2017
Collector(s)	R. T. J. Cappers
Context	–
GIA-accession	40994
Length/Diameter	–
Thickness	–
Weight	–
Quantity	–
Photographer	D. Fennema

Composition

Malus sylvestris	Apple	fruit with exocarp
–	whey	

Description

This qara qurut has been flavoured with pieces of apple and is sold in plastic tubs. It is produced by a company in north-western Iran.

1 cm

1 cm

FIGURE 251

16
Tarhana

Tarhana is a dried cereal-milk product. The grain is fragmented and often par-boiled. This suggests that tarhana would not have become a standard food item until par-boiling of the rachillate spikelets of hulled wheat became a standard procedure during large-scale crop processing. It seems likely that in the first instance only the finest, pulverized fraction of the grain was used for mixing with milk. The modern examples of tarhana show, however, that coarsely ground grain is also used. Basically, two kinds of milk can be used: yoghurt (being fermented milk that contains all of the original nutrients) and skimmed milk or buttermilk (being the low-fat milk that is obtained when fat is removed in butter making). Possibly, tarhana was traditionally made from skimmed milk or buttermilk in order to have access to butter as well.

Tarhana is a food product that shows great variation, which is made possible by mixing the basic ingredients with supplementary ones and by shaping the mixture in different ways. Two subgroups of tarhana can be distinguished based on their shape: a granulated tarhana and a moulded tarhana. The granulated tarhana is produced by pressing the mixture through a sieve. In this way, granules of a similar size are obtained which will all achieve doneness with the same amount of cooking time. The moulded tarhana is produced in all kinds of shapes. The most traditional tarhana is shaped using one or two hands. Such tarhana may show an imprint of parts of the hands. Hand-moulded tarhana can be dried on a layer of aromatic leaves, and large fragments of these leaves may still be present on the bottom of the individual units of tarhana. Tarhana can also be shaped by using moulds, in which uniformly sized pieces are produced.

Tarhana has a pronounced taste, and its consumption as a regular part of the diet will partly depend on the consumer becoming familiar with its taste by regularly eating this food as part of a meal. The fact that it is appreciated as a foodstuff is demonstrated by the supply of homemade tarhana on markets even outside of the production areas.

Tarhana

Obtained	Dordrecht (province of Zuid Holland, the Netherlands)
Origin	Syria
Date	February 2018
Collector(s)	D. Alshawish
Context	Shop
GIA-accession	41147
Length/Diameter	–
Thickness	–
Weight	–
Quantity	–
Photographer	D. Fennema

Composition

Triticum	Wheat	grain kernel
–	salt	
–	yeast	
–	yoghurt	
–	water	

Description

This tarhana is characterized by a powdery consistency. In this case, only the pulverized fraction of the bulgur was used for mixing with the milk product. It is possible that the earliest forms of tarhana involved only this finest fraction of the bulgur and that the coarser granules were initially used for other dishes. The use of yoghurt implies that no fat was extracted from the milk for the production of butter. The trade name is 'local kishk'. The shelf-life according to the packaging is 2 years.

FIGURE 252

Tarhana

Obtained	Amman (Amman governorate, Jordan)
Origin	–
Date	September 2017
Collector(s)	H. Madina
Context	Shop
GIA-accession	41142
Length/Diameter	–
Thickness	–
Weight	–
Quantity	–
Photographer	D. Fennema

Composition

Triticum	Wheat	grain kernel
–	yoghurt	

Description

The tarhana is a mixture of fine-ground bulgur, made of polished grain kernels, and dried yoghurt. The use of yoghurt implies that all nutrients of the milk are present in the tarhana. The milk is probably from sheep. The owner of the shop is from Syria, and the tarhana is sold as 'kishk', a name that is in use in Jordan, Lebanon, Syria, and Egypt.

FIGURE 253

Tarhana

Obtained	Aegina (Attica region, Greece)
Origin	Neos Mylopopus Yiannitsa (Central Macedonia region, Greece)
Date	March 2015
Collector(s)	A. Kokonaki
Context	–
GIA-accession	38098
Length/Diameter	–
Thickness	–
Weight	–
Quantity	–
Photographer	D. Fennema

Composition

Triticum	Wheat	grain kernel
–	non-fermented milk	
–	salt	

Description

This is a factory-made tarhana. Judging from its shape, the cereal–milk mixture was likely first pressed into cylinders and then fragmented into units of irregular length. It is a sweet tarhana, which means that unfermented milk was used. The final boiling time is 8–10 min. The tarhana is made into soup and eaten with feta cheese and rusk bread.

FIGURE 254

Tarhana

Obtained	Almiros (Thessaly region, Greece)
Origin	Velestino (Thessaly region, Greece)
Date	December 2017
Collector(s)	H. R. Reinders & P. de Roever
Context	Market
GIA-accession	41107
Length/Diameter	–
Thickness	–
Weight	–
Quantity	–
Photographer	D. Fennema

Composition

Triticum	Wheat	grain kernel
–	yoghurt	

Description
Home-made tarhana that was bought from the woman who had made it. The particles are rather large and a large proportion of them exhibit one flat face. This indicates that the tarhana was dried in larger units and fragmented after drying. The flat surface has an imprint of the surface into which it was pressed for drying.

FIGURE 255

Tarhana

Obtained	Almiros (Thessaly region, Greece)
Origin	Velestino (Thessaly region, Greece)
Date	December 2017
Collector(s)	H. R. Reinders & P. de Roever
Context	Market
GIA-accession	41106
Length/Diameter	–
Thickness	–
Weight	–
Quantity	–
Photographer	D. Fennema

Composition

Triticum	Wheat	grain kernel
–	non-fermented milk	

Description
Home-made tarhana that was bought from the woman who had made it. The grain kernels are unpolished and medium-ground. The fragments have become partly stuck together. Because non-fermented milk was used, this kind of tarhana is called 'sweet tarhana'.

FIGURE 256

Tarhana

Obtained	Çaybaşi (Ordu province, Turkey)
Origin	–
Date	August 2015
Collector(s)	R.T.J. Cappers & Y. Okur
Context	Private house
GIA-accession	39598
Length/Diameter	–
Thickness	–
Weight	–
Quantity	–
Photographer	D. Fennema

Composition

Anethum graveolens	Dill	fruit (mericarp)
Capsicum annuum	Red pepper	fruit
Lamiaceae	Dead-nettle family	fruit (mericarp)
Lamiaceae	Dead-nettle family	leaf
Ocimum	Basil	fruit (mericarp)
Triticum aestivum ssp. *aestivum*	Bread wheat	grain kernel
–	milk	

Description

The tarhana was obtained from the woman who had made it. A sample of the unprocessed bread wheat was also collected and is stored in the reference collection of the Groningen Institute of Archaeology. The tarhana is sieved in order to obtain small-sized granules. The tarhana is spiced with different plants, including members of the Nightshade family (Solanaceae), the Dead-nettle family (Lamiaceae), and the Carrot family (Apiaceae). Small fragments of these plants, including leaves, seeds, and fruits, are present in the tarhana, whereas the larger fragments are sieved out. It is unknown what kind of milk was used.

FIGURE 257

Tarhana

Obtained	Izmit (Kocaeli province, Turkey)
Origin	–
Date	September 2014
Collector(s)	H. Balcı
Context	–
GIA-accession	34996
Length/Diameter	–
Thickness	–
Weight	–
Quantity	–
Photographer	D. Fennema

Composition

Allium cepa	Onion	bulb
Allium sativum	Garlic	bulb
Capsicum annuum	Red pepper	fruit
Solanum lycopersicum	Tomato	fruit
Triticum aestivum ssp. aestivum	Bread wheat	grain kernel
–	milk	
–	salt	

Description

This kind of granulated tarhana is typical for the Aegean and for the Marmara region of Turkey. The wheat grain is granulated to a fine grade. It is unknown what kind of milk was used. The plant parts used to spice it, including sweet red pepper, onion, garlic, and tomato, are sieved out before the remaining mixture is size-graded with a sieve.

FIGURE 258

Tarhana

Obtained	Hendek (Sakarya province, Turkey)
Origin	Gerede (Bolu province, Turkey)
Date	February 2017
Collector(s)	R. T. J. Cappers & H. Balcı
Context	Shop
GIA-accession	40298
Length/Diameter	–
Thickness	–
Weight	–
Quantity	–
Photographer	D. Fennema

Composition

Allium cepa	Onion	bulb
Mentha	Mint	leaf
Solanum lycopersicum	Tomato	fruit
Triticum aestivum ssp. *aestivum*	Bread wheat	grain kernel
–	salt	
–	yoghurt	

Description

This tarhana is home-made and is produced as a cash crop and sold in a supermarket. The granules differ in size. This tarhana is specially made for babies. Small fragments of the aromatic mint are present in the tarhana. The shelf life is 2 years according to the packaging.

FIGURE 259

Tarhana

Obtained	Balikesir (Balikesir province, Turkey)
Origin	–
Date	August 2014
Collector(s)	C. Cilingir
Context	–
GIA-accession	37163
Length/Diameter	–
Thickness	–
Weight	–
Quantity	–
Photographer	D. Fennema

Composition

Capsicum annuum	Red pepper	fruit
cf. *Lamiaceae*	Dead-nettle family	leaf
Solanum lycopersicum	Tomato	fruit
Triticum turgidum ssp. *durum*	Hard wheat	grain kernel
–	fermented milk	

Description

The tarhana consists of granules of different size. After mixing and shaping it into granules, the tarhana is dried in thin, flat layers. These are crumbled again after drying. Tomato paste is used, and seeds of tomato are present in the tarhana. Small fragments of leaves are present as well, possibly of the Dead-nettle family (Lamiaceae).

FIGURE 260

Tarhana

Obtained	Istanbul (Istanbul province, Turkey)
Origin	Göçkün (Bartin province)
Date	February 2017
Collector(s)	R.T.J. Cappers, M. Ozyarkent & H. Balcı
Context	Farmers market
GIA-accession	40315
Length/Diameter	–
Thickness	–
Weight	–
Quantity	–
Photographer	D. Fennema

Composition

Anethum graveolens	Dill	fruit (mericarp)
Capsicum annuum	Red pepper	fruit
Lamiaceae	Dead-nettle family	fruit (mericarp)
Trigonella foenum-graecum	Fenugreek	leaf
Triticum	Wheat	grain kernel
–	yoghurt	

Description

The tarhana consists of granules that differ in size. The tarhana was made by the mother-in-law of the seller. The tarhana is spiced with members of the Carrot family (Apiaceae), Nightshade family (Solanaceae), Dead-nettle family (Lamiaceae), and Pea family (Fabaceae). The seeds of fenugreek are mixed with the granulated and dried tarhana, whereas the others are added in an earlier stage of preparation.

FIGURE 261

Tarhana

Obtained	Istanbul (Istanbul province, Turkey)
Origin	Erzurum (Erzurum province, Turkey)
Date	August 2017
Collector(s)	B. Alpsoy
Context	–
GIA-accession	40978
Length/Diameter	–
Thickness	–
Weight	–
Quantity	–
Photographer	D. Fennema

Composition

Allium cepa	Onion	bulb
Allium sativum	Garlic	bulb
Anethum graveolens	Dill	fruit (mericarp)
Capsicum annuum	Red pepper	fruit
Cicer arietinum	Chick pea	seed
Cinnamomum	Cinnamon	bark
Curcuma longa	Turmeric	rhizome
Eruca vesicaria	Garden rocket	leaf
Mentha	Mint	leaf
Ocimum basilicum	Basil	leaf
Petroselinum crispum	Parsley	fruit (mericarp)
Piper nigrum	Black pepper	fruit
Solanum lycopersicum	Tomato	fruit
Thymus	Thyme	fruit (mericarp)
Triticum	Wheat	grain kernel
cf. *Urtica*	Nettle	leaf
Zingiber officinale	Ginger	rhizome
–	cow's yoghurt	

Description

The tarhana is home-made. All of the vegetables and herbs are cleaned and finely chopped and mixed with boiled and mashed chick pea, Tomato (small and not too watery), paste of both spicy and mild red and green red pepper, and dried herbs. The wheat flour is added gradually to obtain a smooth, non-grainy mixture. The proportion is 3 kg wheat flour: 4 kg tomatoes: 2 kg onion: 0.7 kg garlic: 3 bundles of fresh herbs (dill, garden rocket, parsley, and mint). The mixture is allowed to ferment in a closed box for 3–4 days, during which period it is stirred a single time. Once it starts to taste a bit sour, the fermentation is considered to be sufficient and the mixture is spread over a clean sheet and sun-dried for 1–2 days. During the drying process, the tarhana is periodically turned to facilitate the drying process. Once dried, the tarhana is granulated with a food processor. Because the tarhana may become a bit humid again during granulating, it is spread out again, this time inside the house. When the tarhana is sufficiently dried, it is put in glass jars for storage.

FIGURE 262

Tarhana

Obtained	Hendek (Sakarya province, Turkey)
Origin	Gerede (Bolu province, Turkey)
Date	February 2017
Collector(s)	R. T. J. Cappers & H. Balcı
Context	Shop
GIA-accession	40301
Length/Diameter	–
Thickness	–
Weight	–
Quantity	–
Photographer	D. Fennema

Composition

Allium cepa	Onion	bulb
Capsicum annuum	Paprika	fruit
Capsicum annuum	Red pepper	fruit
Mentha	Mint	leaf
Solanum lycopersicum	Tomato	fruit
Triticum aestivum ssp. *aestivum*	Bread wheat	grain kernel
–	salt	
–	yoghurt	

Description

This tarhana is home-made and is produced as a cash crop and sold in a supermarket. The granules differ in size. The characteristic orange colour is caused by the incorporation of red-coloured fruits of *Capsicum annuum*, including the paprika and the red pepper. The shelf life is 2 years according to the packaging.

FIGURE 263

Tarhana

Obtained	Hendek (Sakarya province, Turkey)
Origin	Gerede (Bolu province, Turkey)
Date	February 2017
Collector(s)	R. T. J. Cappers & H. Balcı
Context	Shop
GIA-accession	40297
Length/Diameter	–
Thickness	–
Weight	–
Quantity	–
Photographer	D. Fennema

Composition

Allium cepa	Onion	bulb
Cornus mas	Cornelian cherry	fruit
Mentha	Mint	leaf
Solanum lycopersicum	Tomato	fruit
Triticum aestivum ssp. *aestivum*	Bread wheat	grain kernel
–	salt	
–	yoghurt	

Description

This tarhana is home-made and is produced as a cash crop and sold in a supermarket. The granules differ in size. The characteristic pink colour is caused by incorporation of the red fruits of the cornelian cherry. The shelf life is 2 years according to the packaging.

FIGURE 264

Tarhana

Obtained	Istanbul (Istanbul province, Turkey)
Origin	Inebolu (Kastamonu province, Turkey)
Date	February 2017
Collector(s)	R. T. J. Cappers, M. Ozyarkent & H. Balcı
Context	Farmers market
GIA-accession	40314
Length/Diameter	–
Thickness	–
Weight	–
Quantity	–
Photographer	D. Fennema

Composition

Cornus mas	Cornelian cherry	fruit
Triticum	Wheat	grain kernel
–	yoghurt	

Description

This tarhana is called '*kızılcık tarhana*' in Turkish, referring to the cornelian cherry, of which the fruit pulp is added. Small fragments of this fruit are present in the tarhana. The granules clearly differ in size: part of it is almost powdery, whereas part of it consists of rather large conglomerates. It was made by the grandmother of seller. The price was 35 TL/100 g.

FIGURE 265

Tarhana

Obtained	Karaman (Karaman province, Turkey)
Origin	–
Date	August 2013
Collector(s)	Y. Okur
Context	Shop
GIA-accession	32786
Length/Diameter	–
Thickness	–
Weight	–
Quantity	–
Photographer	D. Fennema

Composition

Cicer arietinum	Chick pea	seed
Triticum turgidum ssp. *durum*	Hard wheat	grain kernel
–	sheep's milk	

Description

The solid part of the tarhana bulgur consists of three parts: grain kernels of wheat, seeds of chick pea, and an unidentified part. The grain kernels and seeds of the chick pea are only par-boiled; neither polishing nor grinding has been applied. The tarhana is used for making soup. The final cooking time consists of 16 minutes of heating followed by 15 minutes of leaving it in the hot cooking water. The shelf-life is 1 year according to the packaging.

Nutritional value (100 g)

carbohydrates	62.8 g
lipids	3.1 g
proteins	12.9 g

FIGURE 266

Tarhana

Obtained	Hendek (Sakarya province, Turkey)
Origin	Kahramanmaras (Kahramanmaras province, Turkey)
Date	February 2017
Collector(s)	R.T.J. Cappers & H. Balcı
Context	Shop
GIA-accession	40280
Length/Diameter	–
Thickness	–
Weight	–
Quantity	–
Photographer	D. Fennema

Composition

Lamiaceae	Dead-nettle family	leaf
Triticum	Wheat	grain kernel
–	milk	
–	salt	

Description

The starchy endosperm of the grain kernels has been removed, probably by pressing, during which the lateral sides have split lengthwise. This implies that only the grain envelope, consisting of bran and germ and only a minor fraction of the starchy endosperm, is present. A similar kind of bran is used for the production of hand-shaped tarhana (compare GIA-accessions 40789 and 38971, pages 508 and 574). This tarhana is formed into thin chips and is intended to be eaten as a snack. The thin chips are flattened on a coarse grid, the impressions of which are visible on the final product. Leaves of a member of the Dead-nettle family (Lamiaceae) are added to improve the taste.

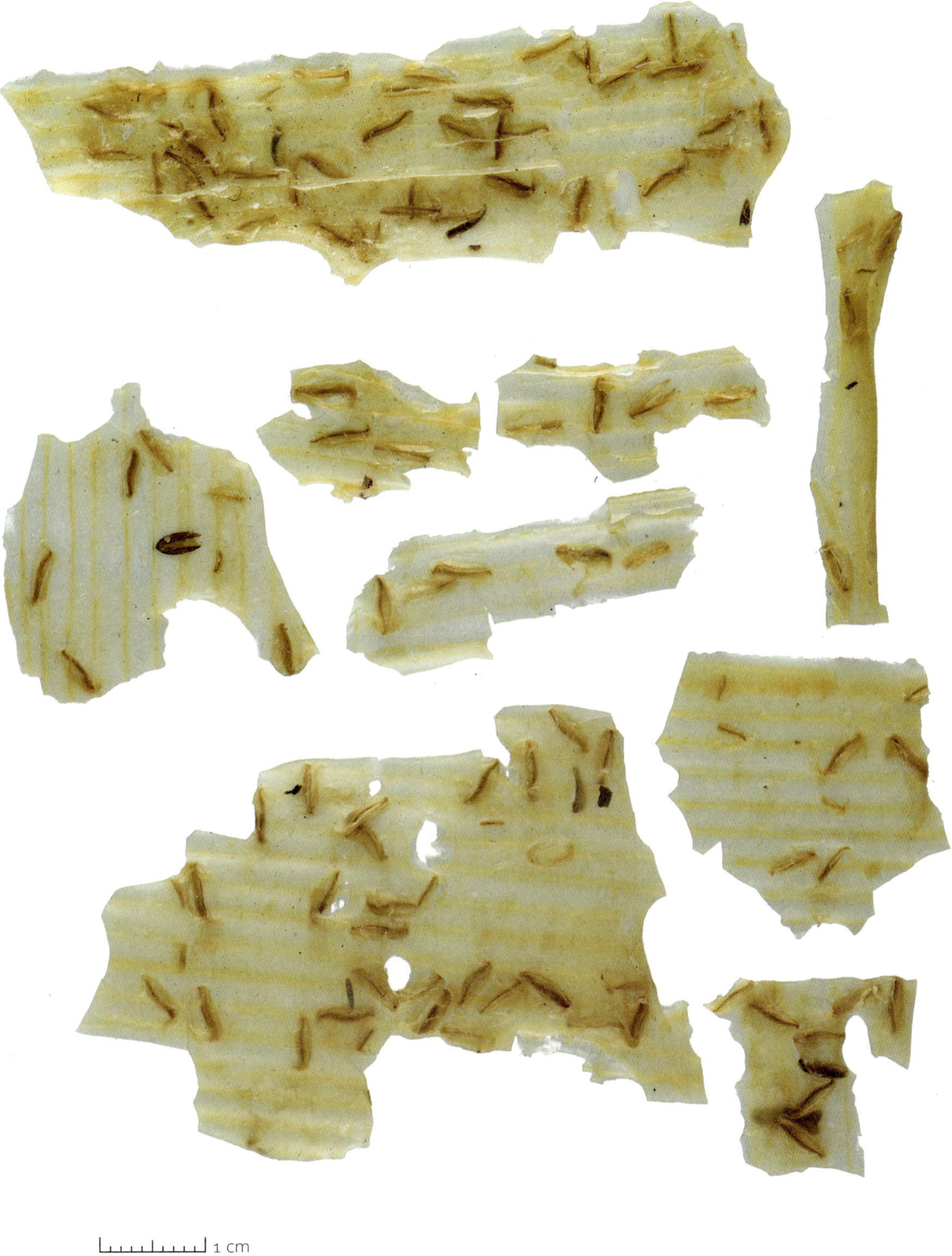

FIGURE 267

Tarhana

Obtained	Karadona (Konya province, Turkey)
Origin	–
Date	May 2015
Collector(s)	R. T. J. Cappers & Y. Okur
Context	Private house
GIA-accession	38969
Length/Diameter	–
Thickness	–
Weight	–
Quantity	–
Photographer	D. Fennema

Composition

Nigella sativa	Black cumin	seed
Triticum	Wheat	grain kernel
–	milk	

Description

The cereal part of this chip tarhana consists of bran fragments, and possibly also germs. This implies that the by-product of polishing was used in the tarhana production. This makes sense in terms of the flat shape of the tarhana. It will improve its nutritional value. A small quantity of seeds of black cumin are added to improve the taste. After mixing and shaping, the tarhana is sliced into rectangular pieces, which measure, on average, 6 × 6 cm. The tarhana chips are eaten as a snack or made into a soup.

1 cm

FIGURE 268

Tarhana

Obtained	Konya (Konya province, Turkey)
Origin	Beysehir (Konya province, Turkey)
Date	May 2004
Collector(s)	R.T.J. Cappers
Context	Shop
GIA-accession	38971
Length/Diameter	16.0–16.6 cm
Thickness	0.2 cm
Weight	(30.1–) 32 (–33.8) g
Quantity	7
Photographer	D. Fennema

Composition

Triticum	Wheat	grain kernel
–	milk	

Description

The starchy endosperm of the grain kernels is removed, probably by pressing, during which the lateral sides split lengthwise. This implies that only the grain envelope, consisting of bran and germ and only a minor fraction of the starchy endosperm, is present. A similar kind of bran is used for the production of hand-shaped tarhana (compare GIA-accessions 40280 and 40789; pages 504 and 574). The tarhana is shaped into flat, circular units and has the appearance of thin chips. The consistent shape, size, and shape suggest that it was industrially produced. It was offered for sale in a plastic bag containing 12 pieces.

FIGURE 269

Tarhana

Obtained	Istanbul (Istanbul province, Turkey)
Origin	Kahramanmaras (Kahramanmaras province, Turkey)
Date	September 2014
Collector(s)	R.T.J. Cappers & Y. Okur
Context	Shop
GIA-accession	34992
Length/Diameter	(6.0–) 7.0 (–8.6) cm
Thickness	(0.7–) 1.1 (–1.8) cm
Weight	(5.8–) 9.6 (–12.6) g
Quantity	13
Photographer	D. Fennema

Composition

Triticum	Wheat	grain kernel
–	milk	

Impurities

Avena	Oat	chaff
Galium	Bedstraw	fruit (mericarp)
Polygonum aviculare	Knotgrass	fruit
Triticum	Wheat	hull
Vaccaria hispanica	Cowherb	seed

Description

The unpolished grain kernels are coarsely ground. The grain is completely embedded in the milk product. Some impurities are, however, visible and these include hull fragments and diaspores of arable weeds. The tarhana pieces are hand-shaped. They have an irregular shape and a light texture. The upper face shows a pattern of irregular, narrow elevations, which does not fit with the imprint of fingers, whereas the bottom face has an imprint of the rather fine, regular grid on which it was dried.

FIGURE 270

Tarhana

Obtained	Katharon (Crete region, Greece)
Origin	–
Date	July 2016
Collector(s)	R.T.J. Cappers
Context	Private house
GIA-accession	39895
Length/Diameter	–
Thickness	–
Weight	–
Quantity	–
Photographer	D. Fennema

Composition

Triticum cf. *turgidum* ssp. *durum*	cf. Hard wheat	grain kernel
–	yoghurt	

Description

The unpolished grain kernels are coarsely ground. The cereal–milk mixture is flattened into a layer c. 2 cm thick that is sun-dried. Once dried, it is crumbled into irregular pieces for storage. The tarhana was obtained from an elderly couple who live in a house outside the village, without electricity, and who are almost self-supporting.

FIGURE 271

Tarhana

Obtained	Kalamafka (Crete region, Greece)
Origin	–
Date	July 2016
Collector(s)	R.T.J. Cappers
Context	Private house
GIA-accession	39896
Length/Diameter	(2.3–) 3.2 (–4.5) cm
Thickness	(1.4–) 2.0 (–3.5) cm
Weight	(1.6–) 3.6 (–6.0) g
Quantity	10
Photographer	D. Fennema

Composition

Triticum aestivum ssp. *aestivum*	Bread wheat	grain kernel
–	yoghurt	

Description

This tarhana was made at our request. A sample of the unprocessed grain was collected for the reference collection of the Groningen Institute of Archaeology. The unpolished grain kernels are coarsely ground. The fermentation process after mixing usually takes 2 days, during which the dough remains in the pan in which the ingredients were mixed. Because we were unable to stay any longer, the fermentation was shortened by 1 day. The tarhana is crumbled by hand in irregular small pieces and sun-dried on the roof of the house. The measurements were obtained on the larger specimens.

FIGURE 272

Tarhana

Obtained	Bulber (Kurdistan province, Iran)
Origin	–
Date	April 2017
Collector(s)	R.T.J. Cappers, M. Kaffashi & P. Hesami
Context	Tourist shop
GIA-accession	40765
Length/Diameter	(9.1–) 10.0 (–10.8) cm
Thickness	(1.3–) 1.4 (–1.6) cm
Weight	(23.7–) 27.0 (–29.8) g
Quantity	4
Photographer	D. Fennema

Composition

Triticum	Wheat	grain kernel
unidentified	–	leaf
–	milk	

Impurities

Hordeum vulgare	Hulled Barley	floret

Description

The unpolished grain kernels are coarsely ground. A small quantity of milk is used. Leaves of an unidentified herb are added to improve the taste. The tarhana is hand-shaped into irregular pieces and has an open texture. This tarhana was bought in a small tourist shop. In addition to typical items for tourists, some local food items were also offered for sale, including small quantities of tarhana and kashk. Other food items that were available include fruits of Sumac (*Rhus coriaria*), whole as well as ground, and fruits of the Almond (*Prunus dulcis*), the Turpentine tree (*Pistacia terebinthus*), the Grape (*Vitis vinifera*), and the Walnut (*Juglans regia*).

FIGURE 273

Tarhana

Obtained	Malekshahi (Ilam province, Iran)
Origin	–
Date	April 2017
Collector(s)	R. T. J. Cappers, M. Kaffashi & P. Hesami
Context	Private house
GIA-accession	40769
Length/Diameter	(4.1–) 5.5 (–6.3) cm
Thickness	(0.8–) 1.0 (–1.3) cm
Weight	(5.6–) 9.5 (–12.6) g
Quantity	19
Photographer	D. Fennema

Composition

Brassica rapa	Turnip	leaf and stem
Capsicum annuum	Red pepper	fruit
Curcuma longa	Turmeric	rhizome
Mentha	Mint	leaf
Triticum aestivum ssp. *aestivum*	Bread wheat	grain kernel
–	milk	
–	spices	

Impurities		
Hordeum vulgare	Hulled Barley	floret
Poaceae	Grass family	fruit
Silene	Campion	seed
Triticum aestivum ssp. *aestivum*	Bread wheat	hull
Triticum aestivum ssp. *aestivum*	Bread wheat	rachis fragment

Description

The unpolished grain kernels are finely ground and mixed with finely chopped turnip. The finest fraction has also been used, giving the tarhana a solid texture. The stem fragments of the turnip are recognizable by their vascular bundles. The tarhana is hand-shaped into small, flat units with an irregular shape. By making small units, quick drying of the tarhana can be ensured. The tarhana is dried on the leaves of mint in order to allow the tarhana to absorb its flavour. The tarhana is made once a year, at the beginning of autumn (November), in large quantities. Part of the tarhana is distributed among family and friends.

FIGURE 274

Tarhana

Obtained	Chenar Bashi (Ilam province, Iran)
Origin	–
Date	April 2017
Collector(s)	R.T.J. Cappers & M. Kaffashi
Context	Private house
GIA-accession	40337
Length/Diameter	(4.4–) 5.9 (–6.8) cm
Thickness	(0.5–) 0.8 (–1.3) cm
Weight	(5.9–) 8.7 (–11.3) g
Quantity	20
Photographer	D. Fennema

Composition

Capsicum annuum	Red pepper	fruit
Curcuma longa	Turmeric	rhizome
Lamiaceae	Dead-nettle family	whole plant
Origanum	Oregano	leaf
Triticum	Wheat	
–	milk	
–	salt	

Impurities

Elytrigia repens	Common couch	floret
Hordeum vulgare	Hulled Barley	floret
Triticum	Wheat	culm fragment
Triticum	Wheat	hulls

Description

The unpolished grain kernels are coarsely ground, and whole grain kernels are present as well. The kernels are mixed with a finely chopped member of the Dead-nettle family. The tarhana is hand-shaped into a flat layer and fragmented into somewhat irregular-shaped units. The tarhana is dried on leaves of the oregano, and large fragments of this plant are still stuck to the bottom face. This tarhana was obtained from the shop owner. When we asked for tarhana in the shop, the shop owner walked to his house, where he collected a quantity. The tarhana had been made by his wife 2–3 months earlier.

FIGURE 275

Tarhana

Obtained	Nevin (Kurdistan province, Iran)
Origin	–
Date	April 2017
Collector(s)	R.T.J. Cappers & M. Kaffashi
Context	Private house
GIA-accession	40771
Length/Diameter	8.8–10.1 cm
Thickness	0.9–1.1 cm
Weight	25.6–33.7 g
Quantity	2
Photographer	D. Fennema

Composition

Brassica rapa	Turnip	leaf and stem
Triticum aestivum ssp. *aestivum*	Bread wheat	grain kernel
–	milk	

Description

The unpolished grain kernels are coarsely ground and mixed with finely chopped leaves and stems of turnip. The tarhana is hand-shaped into flat, round units. The leaves have become somewhat blackish, possibly because they were dried in the sun rather than in the shade. The stem fragments are recognizable by their vascular bundles. The tarhana has an open texture and an irregular surface.

FIGURE 276

Tarhana

Obtained	Sanandaj (Kurdistan province, Iran)
Origin	–
Date	April 2017
Collector(s)	R. T. J. Cappers & M. Kaffashi
Context	Bazaar
GIA-accession	40774
Length/Diameter	(7.1–) 7.5 (–7.9) cm
Thickness	(1.6–) 2.0 (–2.7) cm
Weight	(30.7–) 34.1 (–36.8) g
Quantity	2
Photographer	D. Fennema

Composition

Brassica rapa	Turnip	leaf and stem
Triticum	Wheat	grain kernel
–	milk	

Description

The unpolished grain kernels are coarsely ground, although many grains are still whole. The grain is mixed with chopped turnip. The stem fragments are recognizable by their vascular bundles. The tarhana is hand-shaped into units that vary in shape and size and that have an open texture.

FIGURE 277

Tarhana

Obtained	Sirvan (Ilam province, Iran)
Origin	–
Date	April 2017
Collector(s)	R. T. J. Cappers & M. Kaffashi
Context	Farmhouse
GIA-accession	41002
Length/Diameter	7.5–8.0 cm
Thickness	1.2–1.7 cm
Weight	43.9–48.4 g
Quantity	2
Photographer	D. Fennema

Composition

Foeniculum vulgare	Fennel	fruit (mericarp)
Mentha	Mint	whole plant
Triticum aestivum ssp. *aestivum*	Bread wheat	grain kernel
–	milk	

Impurities		
Avena	Oat	fruit
Hordeum vulgare	Hulled Barley	floret
Triticum aestivum ssp. *aestivum*	Bread wheat	hulls
Vaccaria hispanica	Cowherb	seed

Description

The unpolished grain kernels are coarsely ground. The tarhana is hand-shaped, but imprints of the fingers are not present. The taste is improved by adding fragments of mint and fruits of fennel. Most fruits of fennel have lost their fruit wall, and only seeds are visible. Because the aromatic components are present in the oil bodies of the fruit, it is assumed that these parts are still present.

FIGURE 278

Tarhana

Obtained	Ilam (Ilam province, Iran)
Origin	–
Date	April 2017
Collector(s)	R.T.J.Cappers & M.Kaffashi
Context	Private house
GIA-accession	40752
Length/Diameter	(7.0–) 7.2 (–7.6) cm
Thickness	(0.7–) 0.8 (–0.9) cm
Weight	(24.7–) 27.2 (–30.8) g
Quantity	3
Photographer	D.Fennema

Composition

Brassica rapa	Turnip	leaf
Carum carvi	Caraway	fruit (mericarp)
Foeniculum vulgare	Fennel	fruit (mericarp)
Solanum lycopersicum	Tomato juice	fruit
Triticum aestivum ssp. *aestivum*	Bread wheat	grain kernel
–	buttermilk	
–	salt	
–	water	

Impurities

Triticum aestivum ssp. *aestivum*	Bread wheat	hulls

Description

The unpolished grain kernels are coarsely ground. The tarhana is hand-shaped into flat, round units. Imprints of the fingers are visible on the upper face. The tarhana was made by the mother of M.Kaffashi in September 2016. A total of 70 tarhana pieces were made in a single session.

FIGURE 279

Tarhana

Obtained	Ilam (Ilam province, Iran)
Origin	–
Date	April 2017
Collector(s)	R. T. J. Cappers & M. Kaffashi
Context	Shop
GIA-accession	40763
Length/Diameter	(6.6–) 7.4 (–8.1) cm
Thickness	(1.4–) 1.6 (–1.8) cm
Weight	(27.6–) 39.0 (–46.1) g
Quantity	3
Photographer	D. Fennema

Composition

Brassica rapa ssp. *rapa*	Turnip	leaf
Capsicum annuum	Red pepper	fruit
Curcuma longa	Turmeric	rhizome
Foeniculum vulgare	Fennel	fruit (mericarp)
Mentha	Mint	whole plant
Satureja	Savory	whole plant
Solanum lycopersicum	Tomato	fruit
Triticum aestivum ssp. *aestivum*	Bread wheat	grain kernel
	buttermilk	
–	curry powder	

Description

The unpolished grain kernels are coarsely ground. This tarhana is designated as '*tarhana shalmineh*', referring to the inclusion of turnip. The tarhana was made by the wife of the shop owner and sold in the shop. The tarhana is hand-moulded on a flat work surface. The imprint of fingers is unclear. The bottom shows a high concentration of fragments of turnip and mint, suggesting that these ingredients are only partly mixed in and therefore present on the work surface against which they are flattened. The ingredients are a mixture of indigenous plants and exotic plants. The latter group is represented by turmeric (India, Sri Lanka, south-east Asia), tomato, and red pepper (both New World).

FIGURE 280

Tarhana

Obtained	Ilam (Ilam province, Iran)
Origin	Malekshahi (Ilam province, Iran)
Date	April 2017
Collector(s)	R. T. J. Cappers & M. Kaffashi
Context	Shop
GIA-accession	40756
Length/Diameter	(8.5–) 9.3 (–10.6) cm
Thickness	(1.3–) 1.4 (–1.5) cm
Weight	(45.7–) 55.4 (–66.6) g
Quantity	3
Photographer	D. Fennema

Composition

cf. *Capsicum annuum*	Red pepper	fruit
Foeniculum vulgare	Fennel	fruit (mericarp)
Triticum aestivum ssp. *aestivum*	Bread wheat	grain kernel
–	milk	

Impurities

Avena	Oat	fruit
Hordeum vulgare	Hulled Barley	floret
Triticum aestivum ssp. *aestivum*	Bread wheat	hulls

Description

The polished grain kernels are coarsely ground. The tarhana is hand-shaped into pieces that vary in size. Imprints of the fingers are present on the upper face, whereas the bottom face is rather flat. The tarhana is spiced with fennel and possibly red pepper. Impurities consist of hulled barley and wild oat.

FIGURE 281

Tarhana

Obtained	Ilam (Ilam province, Iran)
Origin	–
Date	April 2017
Collector(s)	R. T. J. Cappers & M. Kaffashi
Context	Shop
GIA-accession	40753
Length/Diameter	7.9–8.6 cm
Thickness	1.8–1.9 cm
Weight	33.9–36.2 g
Quantity	2
Photographer	D. Fennema

Composition

Carum carvi	Caraway	fruit (mericarp)
Curcuma longa	Turmeric	rhizome
Foeniculum vulgare	Fennel	fruit (mericarp)
Triticum	Wheat	grain kernel
–	buttermilk	
–	salt	

Description

A mixture of coarse- and fine-ground bulgur was used for this tarhana. The tarhana is hand-shaped into flat, oval units without a clear imprint of fingers on either face. The outside of the tarhana is covered with a milky layer that prevents a further visual analysis of its composition. The flour improves the consistency of the tarhana soup. The original colour of the tarhana was yellow due to the use of turmeric. This colour becomes visible at a fresh fracture. The tarhana was made by wife of the shop owner.

FIGURE 282

Tarhana

Obtained	Sirvan (Ilam province, Iran)
Origin	–
Date	April 2017
Collector(s)	R.T.J. Cappers & M. Kaffashi
Context	Farmhouse
GIA-accession	40782
Length/Diameter	10.8 cm
Thickness	1.6 cm
Weight	98.1 g
Quantity	1
Photographer	D. Fennema

Composition

Brassica rapa	Turnip	leaf
Foeniculum vulgare	Fennel	fruit (mericarp)
Triticum aestivum ssp. *aestivum*	Bread wheat	grain kernel
–	milk	

Impurities

Hordeum vulgare	Hulled Barley	floret
Vaccaria hispanica	Cowherb	seed

Description

The unpolished grain kernels are coarsely ground. Most grain kernels are still intact. The tarhana is hand-shaped into a thick, round unit. This tarhana is designated as '*tarhana shalmineh*', referring to the inclusion of turnip. In addition, fruits of fennel are added to improve its taste. Impurities consist of barley and a diaspore of an arable weed.

FIGURE 283

Tarhana

Obtained	Sanandaj (Kurdistan province, Iran)
Origin	Marivan (Kurdistan province, Iran)
Date	April 2017
Collector(s)	R. T. J. Cappers & M. Kaffashi
Context	Bazaar
GIA-accession	40783
Length/Diameter	13.4 cm
Thickness	2.0 cm
Weight	94.2 g
Quantity	1
Photographer	D. Fennema

Composition

Brassica rapa	Turnip	leaf and stem
Triticum aestivum ssp. *aestivum*	Bread wheat	grain kernel
–	milk	

Description

The unpolished grain kernels are coarsely ground, with many grain kernels still intact. This tarhana is designated as '*tarhana shalmineh*', referring to the inclusion of turnip. The tarhana is hand-shaped into rather large units. The bottom face is flat, whereas the upper face varies from flat to more convex. The stem fragments are recognizable by their vascular bundles. The tarhana has an open texture and an irregular surface.

FIGURE 284

Tarhana

Obtained	Hendek (Sakarya province, Turkey)
Origin	Kahramanmaras (Kahramanmaras province, Turkey)
Date	February 2017
Collector(s)	R. T. J. Cappers & H. Balcı
Context	Shop
GIA-accession	40277
Length/Diameter	7.0–7.5 cm
Thickness	0.8–1.5 cm
Weight	31.9–46.2 g
Quantity	–
Photographer	D. Fennema

Composition

Triticum turgidum ssp. *durum*	Hard wheat	grain kernel
–	milk	

Description

The polished grain kernels are coarsely ground. The bottom face, and to a lesser extent also the upper face, is covered with a white powder. Its composition has not yet been identified, but most probably it is a mixture of flour and salt. This could prevent infestation by insects. The tarhana is home-made and was sold in a shop. Most pieces have become fragmented. The tarhana pieces are rather solid and heavy. The shelf life is 1 year according to an advertisement in the shop.

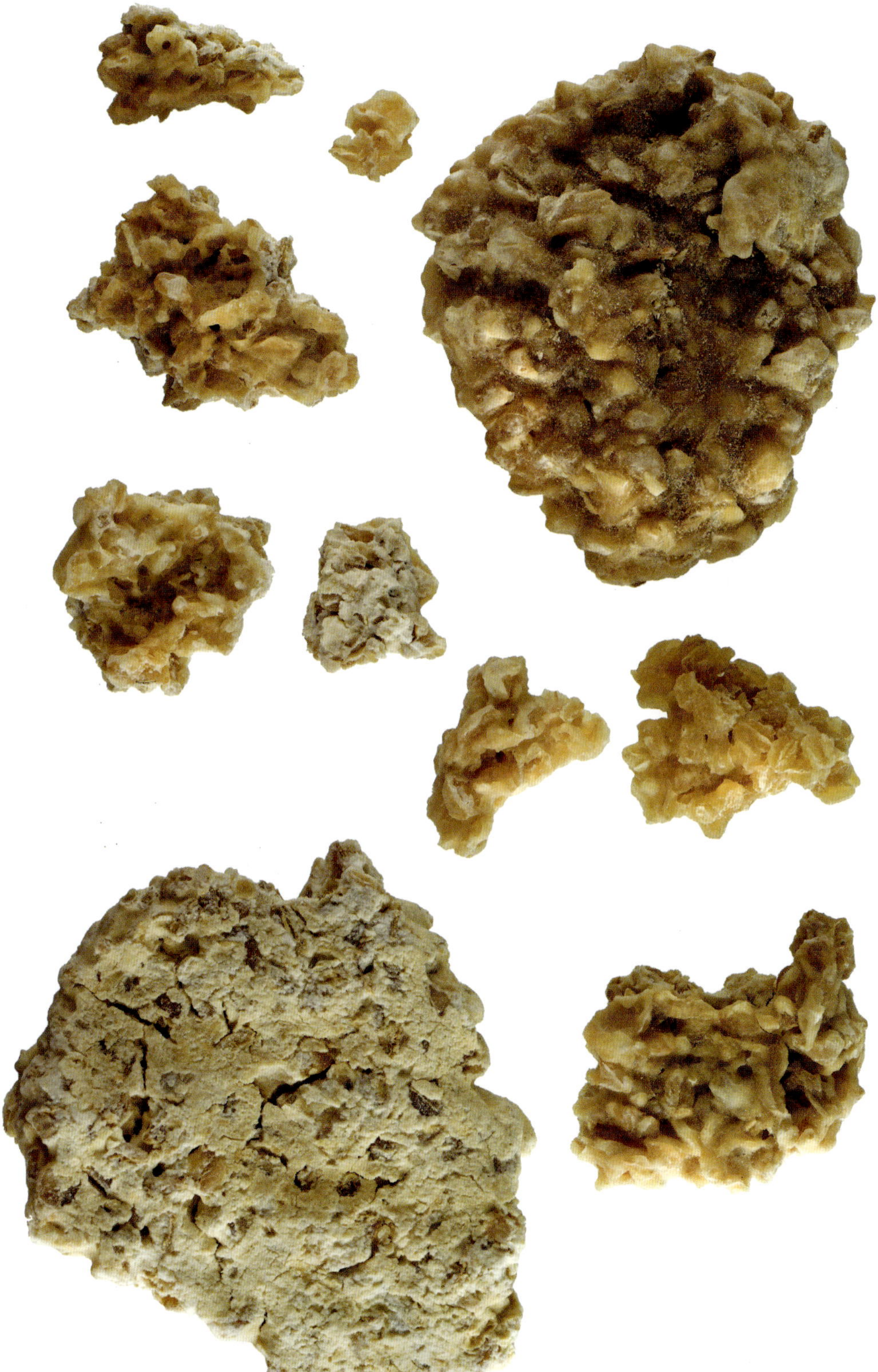

FIGURE 285

Tarhana

Obtained	Sanandaj (Kurdistan province, Iran)
Origin	–
Date	April 2017
Collector(s)	R.T.J. Cappers & M. Kaffashi
Context	Bazaar
GIA-accession	40780
Length/Diameter	5.9–6.4 cm
Thickness	1.3–1.4 cm
Weight	16.7–19.6 g
Quantity	2
Photographer	D. Fennema

Composition

Triticum	Wheat	grain kernel
–	milk	
–	salt	

Description

The unpolished grain kernels are coarsely ground. The tarhana is hand-shaped into rather small and somewhat elongated, flat units. The texture is open. A white salty and powdery substance is present on the surface, especially on the bottom face, but also within the tarhana units.

FIGURE 286

Tarhana

Obtained	Tehran (Tehran province, Iran)
Origin	–
Date	April 2017
Collector(s)	R. T. J. Cappers & M. Kaffashi
Context	Shop
GIA-accession	40741
Length/Diameter	8.6 cm
Thickness	2.8 cm
Weight	49.5 g
Quantity	1
Photographer	D. Fennema

Composition

Brassica rapa	Turnip	leaf
Triticum	Wheat	grain kernel
–	milk	

Description

The unpolished grain kernels are finely ground. The tarhana is shaped into elongated, rather thick units using two hands, resulting in the imprint of the fingertips from one hand and the flat shape of the palm of the other hand. Although the texture is rather solid, the unit has some deep cracks. Ground but unboiled grain has stuck to the bottom of the units, indicating that unboiled grain was used to isolate the tarhana from the drying surface.

FIGURE 287

Tarhana

Obtained	Göcer (Karaman province, Turkey)
Origin	–
Date	August 2014
Collector(s)	Y. Okur
Context	Private house
GIA-accession	40976
Length/Diameter	(6.1–) 7.3 (–8.6) cm
Thickness	(0.7–) 1.0 (–1.4) cm
Weight	(16.4–) 24.7 (–38.6) g
Quantity	12
Photographer	D. Fennema

Composition

Triticum aestivum ssp. *aestivum*	Bread wheat	grain kernel
–	salt	
–	yoghurt	

Description

The unpolished grain kernels are coarsely ground. The process, according to the woman who made this tarhana, is: 1) boiling grain, 2) drying grain, 3) granulating grain, 4) mixing grain and yoghurt, 5) shaping the tarhana, 6) drying tarhana. The tarhana is hand-shaped. An enquiry about the function of the imprint of a fingertip in the middle of the upper face did not yield a specific reason. Possibly, this imprint is related to the possibility of piercing the tarhana pieces in order to string them for safe storage. If so, the tradition of shaping is being continued without a clear understanding of its original function. The tarhana is home-made. It was made in Turkey and brought to Amsterdam as a food supply.

FIGURE 288

Tarhana

Obtained	Sanandaj (Kurdistan province, Iran)
Origin	–
Date	April 2017
Collector(s)	R.T.J. Cappers & M. Kaffashi
Context	Bazaar
GIA-accession	40770
Length/Diameter	8.6–9.5 cm
Thickness	0.8–0.9 cm
Weight	45.0–46.5 g
Quantity	2
Photographer	D. Fennema

Composition

Triticum aestivum ssp. *aestivum*	Bread wheat	grain kernel
–	sheep's buttermilk	

Description

The unpolished grain kernels are coarsely ground. The tarhana is hand-shaped into flat, round units. Imprints of the fingers are visible on the top face of the units. The texture is rather open. The shop owner mentioned that these specimens are small and that bigger ones are also made.

FIGURE 289

Tarhana

Obtained	Sanandaj (Kurdistan province, Iran)
Origin	Marivan (Kurdistan province, Iran)
Date	April 2017
Collector(s)	R.T.J. Cappers & M. Kaffashi
Context	Bazaar
GIA-accession	40766
Length/Diameter	14.1 cm
Thickness	1.8 cm
Weight	187.5 g
Quantity	1
Photographer	D. Fennema

Composition

Triticum aestivum ssp. *aestivum*	Bread wheat	grain kernel
–	milk	

Impurities

Hordeum vulgare	Hulled Barley	floret
Triticum aestivum ssp. *aestivum*	Bread wheat	hulls

Description

The unpolished grain kernels are coarsely ground. Whole grain kernels are present as well. The tarhana is hand-shaped in relatively large, circular units, showing the imprint of entire fingers. This tarhana is designated as '*tarhana dooineh*', which translates as '*tarhanan dough*', a term that indicates plain tarhana, meaning that no turnip was used.

FIGURE 290

Tarhana

Obtained	Ilam (Ilam province, Iran)
Origin	Shrivan (Ilam province, Iran)
Date	April 2017
Collector(s)	R.T.J. Cappers & M. Kaffashi
Context	Shop
GIA-accession	40759
Length/Diameter	(9.8–) 10.1 (–10.5) cm
Thickness	(0.5–) 0.7 (–0.8) cm
Weight	(40.3–) 46.2 (–51.2) g
Quantity	3
Photographer	D. Fennema

Composition

Apiaceae	Carrot family	fruit (mericarp)
Brassica rapa	Turnip	leaf
Lamiaceae	Dead-nettle family	whole plant
Triticum aestivum ssp. *aestivum*	Bread wheat	grain kernel
–	milk	

Description

The unpolished grain kernels are finely ground, but some grain kernels remain intact. The ground grain is mixed with spices, of which only small parts are visible at the surface. The tarhana is hand-shaped into thin, flat, round units. The upper face shows imprints of fingers running in different directions, indicating that the final shaping is done by flattening the tarhana units against a work surface. This tarhana is designated as '*tarhana shalmineh*', a term that refers to the use of turnip. The tarhana is dried on turnip leaves, and fragments of the leaves have become attached to the bottom of the units, giving them a characteristic design. Some leaves exhibit the roughly fringed ends of the stem, indicating that the leaves were removed by pulling them downwards.

FIGURE 291

Tarhana

Obtained	Jafar Abad (Ilam province, Iran)
Origin	–
Date	April 2017
Collector(s)	R.T.J. Cappers & M. Kaffashi
Context	Private house
GIA-accession	40755
Length/Diameter	(9.4–) 10.5 (–12.5) cm
Thickness	(0.7–) 1.1 (–1.5) cm
Weight	(29.3–) 45.9 (–57.5) g
Quantity	5
Photographer	D. Fennema

Composition

Curcuma longa	Turmeric	rhizome
Foeniculum vulgare	Fennel	fruit (mericarp)
Triticum aestivum ssp. *aestivum*	Bread wheat	grain kernel
–	milk	

Impurities

Cephalaria syriaca	–	fruit and perianth
Poaceae	Grass family	fruit

Description

The tarhana is hand-shaped into flat, irregular, and somewhat undulating units. It has deep imprints on the bottom of the screen on which it was dried. The upper face shows the imprint of fingers. The tarhana was given to us by a shop owner, even though tarhana is not sold in his shop. When we asked for some tarhana, he walked home to collect some for us.

FIGURE 292

Tarhana

Obtained	Ilam (Ilam province, Iran)
Origin	–
Date	April 2017
Collector(s)	R. T. J. Cappers & M. Kaffashi
Context	Shop
GIA-accession	40761
Length/Diameter	c. 11 cm
Thickness	c. 9 cm
Weight	–
Quantity	–
Photographer	D. Fennema

Composition

Foeniculum vulgare	Fennel	fruit (mericarp)
Lens culinaris	Lentil	seed
Mentha	Mint	whole plant
cf. *Origanum*	Oregano	leaf
Solanum lycopersicum	Tomato	fruit
Triticum aestivum ssp. *aestivum*	Bread wheat	grain kernel
–	milk	

Impurities

Daucus carota	Carrot	fruit (schizocarp)
Hordeum vulgare	Hulled Barley	floret
Triticum aestivum ssp. *aestivum*	Bread wheat	hull
Triticum aestivum ssp. *aestivum*	Bread wheat	rachis fragment

Description

The unpolished grain kernels are coarsely ground. The wheat is designated as '*horoosh*', which is a special kind of wheat used for making tarhana. The tarhana is hand-shaped, and the upper face shows the imprint of fingers. The tarhana is dried on a layer of aromatic leaves from members of the Dead-nettle family. These leaves, and even fragments of the stem, become attached to the bottom of the tarhana units. The tarhana was made by the wife of the shop owner. Only fragments could be collected, and therefore the original length/diameter of the units has been estimated.

FIGURE 293

Tarhana

Obtained	Ganjeh (Ilam province, Iran)
Origin	–
Date	April 2017
Collector(s)	R.T.J. Cappers & M. Kaffashi
Context	Farmhouse
GIA-accession	40781
Length/Diameter	8.5–10.5 cm
Thickness	0.8–0.9 cm
Weight	21.3–27.6 g
Quantity	2
Photographer	D. Fennema

Composition

Capsicum annuum	Red pepper	fruit
Curcuma longa	Turmeric	rhizome
Foeniculum vulgare	Fennel	fruit (mericarp)
Origanum	Oregano	leaf
Triticum aestivum ssp. *aestivum*	Bread wheat	grain kernel
–	buttermilk	
–	salt	

Impurities

Triticum aestivum ssp. *aestivum*	Bread wheat	hulls

Description

The unpolished grain kernels are coarsely ground. The tarhana is hand-shaped into flat, irregular units. Imprints of fingers are visible on the upper face. Remains of the oregano are visible on the bottom face, indicating that the tarhana was dried on a bed of this aromatic plant. This tarhana is designated as '*tarhana dooineh*', a term that indicates that no turnip was used. It was obtained from a family that lives rather isolated in the mountains in the Canyon of Razianeh.

FIGURE 294

Tarhana

Obtained	Sanandaj (Kurdistan province, Iran)
Origin	–
Date	April 2017
Collector(s)	R.T.J. Cappers & M. Kaffashi
Context	Bazaar
GIA-accession	40772
Length/Diameter	(7.9–) 8.0 (–8.1) cm
Thickness	(2.6–) 2.8 (–2.9) cm
Weight	(82.1–) 88.7 (–95.2) g
Quantity	2
Photographer	D. Fennema

Composition

Triticum	Wheat	grain kernel
–	milk	
–	salt	

Impurities

Poaceae	Grass family	floret

Description

The unpolished grain kernels are coarsely ground. The tarhana is hand-shaped into round units that are relatively thick. The grain fragments are coated with a salty tasting white powder, which forms an almost continuous layer on the bottom face.

FIGURE 295

Tarhana

Obtained	Ilam (Ilam province, Iran)
Origin	Salab Karazan (Ilam province, Iran)
Date	April 2017
Collector(s)	R.T.J. Cappers & M. Kaffashi
Context	Shop
GIA-accession	40760
Length/Diameter	(6.5–) 6.7 (–7.0) cm
Thickness	(1.7–) 1.9 (–2.1) cm
Weight	(32.3–) 34.1 (–37.5) g
Quantity	3
Photographer	D. Fennema

Composition

Brassica rapa	Turnip	leaf
Capsicum annuum	Red pepper	fruit
Carum carvi	Caraway	fruit (mericarp)
Foeniculum vulgare	Fennel	fruit (mericarp)
Lamiaceae	Dead-nettle family	whole plant
Triticum	Wheat	grain kernel
–	milk	

Impurities

Rumex	Dock	fruit
Triticum	Wheat	hull

Description

The unpolished grain kernels are coarsely ground and mixed with chopped aromatic plants. This tarhana is designated as '*tarhana shalmineh*', a term that refers to the use of turnip. The tarhana was made by the wife of the shop owner, and its composition was related to us by the shop owner. The tarhana is hand-shaped into flat, round units. The bottom face has thin, elongated imprints, possibly of a grid on which the tarhana was dried.

FIGURE 296

Tarhana

Obtained	Sanandaj (Kurdistan province, Iran)
Origin	Negel (Kurdistan province, Iran)
Date	April 2017
Collector(s)	R. T. J. Cappers & M. Kaffashi
Context	Bazaar
GIA-accession	40775
Length/Diameter	8.5–9.6 cm
Thickness	1.3–1.5 cm
Weight	35.5–51.5 g
Quantity	2
Photographer	D. Fennema

Composition

Brassica rapa	Turnip	leaf
Triticum	Wheat	grain kernel
Vitis vinifera	Grape	fruit

Impurities

Triticum	Wheat	rachis fragment
–	insect	

Description

The unpolished grain kernels are coarsely ground. The tarhana is hand-shaped into flat, elongated units. A shallow imprint of fingers is visible. This tarhana is designated as '*tarhana shalmineh*', a term that refers to the use of turnip. The characteristic purple colour is caused by the use of grape fruits. A seed of grape is visible at the surface. Although the tarhana was frozen for a week at −40 °C, an insect was found after the tarhana was removed from the freezer that must have developed from an infestation that occurred during storage.

FIGURE 297

Tarhana

Obtained	Sanandaj (Kurdistan province, Iran)
Origin	–
Date	April 2017
Collector(s)	R. T. J. Cappers & M. Kaffashi
Context	Bazaar
GIA-accession	40776
Length/Diameter	10.4–10.9 cm
Thickness	1.3 cm
Weight	81.1–84.4 g
Quantity	2
Photographer	D. Fennema

Composition

Triticum aestivum ssp. *aestivum*	Bread wheat	grain kernel
Vitis vinifera	Grape	fruit
–	milk	

Impurities

Avena	Oat	floret
Hordeum vulgare	Hulled Barley	floret
Triticum aestivum ssp. *aestivum*	Bread wheat	culm fragment
Triticum aestivum ssp. *aestivum*	Bread wheat	hulls

Description

The unpolished grain kernels are coarsely ground, although many grain kernels have not fragmented. The grain is mixed with fruits of grape, giving the tarhana its characteristic purple colour. The tarhana is hand-shaped into flat, round units. Imprints of fingers are present on the upper face.

FIGURE 298

Tarhana

Obtained	Tehran (Tehran province, Iran)
Origin	Sanandaj (Kurdistan province, Iran)
Date	July 2017
Collector(s)	R. T. J. Cappers & M. Kaffashi
Context	Private house
GIA-accession	40861
Length/Diameter	9.8 cm
Thickness	1.7 cm
Weight	46.4 g
Quantity	1
Photographer	D. Fennema

Composition

Triticum turgidum ssp. *durum*	Hard wheat	grain kernel
–	yoghurt	

Description

The unpolished grain kernels are coarsely ground, although some of the grain kernels have not fragmented. The tarhana is hand-shaped and fits in the palm. Imprints of the fingers are barely visible due to the coarse structure. The piece of tarhana was obtained from a woman living in Tehran. It was made by her mother, who lives in Sanandaj.

FIGURE 299

Tarhana

Obtained	Sanandaj (Kurdistan province, Iran)
Origin	Javanrud (Kurdistan province, Iran)
Date	April 2017
Collector(s)	R. T. J. Cappers & M. Kaffashi
Context	Bazaar
GIA-accession	40768
Length/Diameter	19.8 cm
Thickness	1.6 cm
Weight	225 g
Quantity	1
Photographer	D. Fennema

Composition

Foeniculum vulgare	Fennel	fruit (mericarp)
Triticum	Wheat	grain kernel
–	cow's buttermilk	

Description

The hand-shaped tarhana is modelled into large, elongated units. The tarhana easily breaks into smaller pieces due to open texture and the large size of the units. One might wonder to what extent intact units are stored. The milk product concerns buttermilk of a cow. The use of buttermilk entails that the tarhana has a low fat content. The tarhana is spiced with fennel, but the quantity of fennel fruits seems to be low.

FIGURE 300

Tarhana

Obtained	Karadona (Konya province, Turkey)
Origin	–
Date	May 2015
Collector(s)	R. T. J. Cappers & Y. Okur
Context	Private house
GIA-accession	39361
Length/Diameter	(5.0–) 5.7 (–7.0) cm
Thickness	(1.1–) 1.5 (–2.3) cm
Weight	(4.2–) 6.0 (–7.0) g
Quantity	7
Photographer	D. Fennema

Composition

Triticum	Wheat	grain kernel
–	milk	

Description

The unpolished grain kernels are coarsely ground. The tarhana is shaped in the hand, and finger prints are visible on the upper face.

FIGURE 301

Tarhana

Obtained	Amsterdam (province of Noord Holland, the Netherlands)
Origin	Kahramanmaras (Kahramanmaras province, Turkey)
Date	May 2017
Collector(s)	Y. Okur
Context	Shop
GIA-accession	40789
Length/Diameter	(7.1–) 8.2 (–9.4) cm
Thickness	(1.8–) 2.1 (–2.9) cm
Weight	(26.3–) 31.8 (–39.3) g
Quantity	30
Photographer	D. Fennema

Composition

Thymus	Thyme	whole plant
Triticum aestivum ssp. *aestivum*	Bread wheat	grain kernel
–	salt	
–	yoghurt (low in fat)	

Impurities

Poaceae	Grass family	floret
Ranunculus arvensis	Corn buttercup	fruit
Triticum aestivum ssp. *aestivum*	Bread wheat	culm fragment
Triticum aestivum ssp. *aestivum*	Bread wheat	hulls

Description

Contrary to what is mentioned on the packaging, this tarhana is made with grain of bread wheat (i.e. naked wheat), rather than a hulled wheat, as indicated by the presence of rachis fragments of bread wheat and whole grain kernels. The starchy endosperm is removed from the grain kernels. It is the remaining bran of the whole grain kernels that is used for the production of the tarhana. A similar kind of bran is used for the production of chips (compare GIA-accessions 40280 and 38971; pages 504 and 508). The thyme is finely ground. The tarhana units are shaped using both hands, as evidenced by imprints of the palm of one hand and 3–4 fingers of the other hand. The tarhana has an open texture. Impurities include small fragments of hulls and, occasionally, diaspores of arable weeds. Although the tarhana is hand-shaped, it seems to be produced from industrially produced ingredients. The tarhana was bought in a Turkish shop, where it was sold in 1 kg bags. The shelf life according to the packaging is 1.5 years.

Nutritional value (100 g)

carbohydrates	22 g
lipid	3 g
proteins	5 g

FIGURE 302

Tarhana

Obtained	Thomadiano (Crete region, Greece)
Origin	–
Date	May 2014
Collector(s)	R. T. J. Cappers
Context	Private house
GIA-accession	34741
Length/Diameter	(4.3–) 5.4 (–6.6) cm
Thickness	(1.6–) 1.9 (–2.1) cm
Weight	(8.5–) 9.6 (–10.3) g
Quantity	5
Photographer	D. Fennema

Composition

Triticum aestivum ssp. *aestivum*	Bread wheat	grain kernel
–	milk	

Description

The unpolished grain kernels coarsely ground. The tarhana is hand-shaped into elongated and irregular-shaped pieces. Imprints of fingers are present but not clearly visible.

FIGURE 303

Tarhana

Obtained	Cairo (Cairo governorate, Egypt)
Origin	Beni Suef (Beni Suef governorate, Egypt)
Date	August 2017
Collector(s)	A. Shahat
Context	Street vendor
GIA-accession	40991
Length/Diameter	(4.0–) 5.1 (–5.7) cm
Thickness	(1.1–) 2.7 (–3.4) cm
Weight	(8.4–) 18.1 (–23.6) g
Quantity	8
Photographer	D. Fennema

Composition

Triticum aestivum ssp. *aestivum*	Bread wheat	grain kernel
–	fermented milk	
Impurities		
Hordeum vulgare	Hulled Barley	floret
Phalaris	Canary-grass	floret
–	insect	

Description

This tarhana was sold to us by a woman on the street. The processing that is applied by her consists of the following steps. First, the unpolished grain kernels are par-boiled and then sun-dried. After grinding, they are mixed with some water, and then sour milk is added. The tarhana is hand-shaped into irregular humps with a flat bottom face. The final cooking procedure starts with soaking the tarhana in some water for 5 minutes. Then the tarhana pieces are squeezed. They are further cooked, eggs are added, and then the mixture is served as a dish. The tarhana is infested with beetles, an infestation which must have occurred during storage. Other impurities include barley and a diaspore of an arable weed.

FIGURE 304

Tarhana

Obtained	Fayum (Fayum governorate, Egypt)
Origin	–
Date	November 2006
Collector(s)	R.T.J. Cappers
Context	Market
GIA-accession	29124
Length/Diameter	(4.1–) 4.4 (–4.5) cm
Thickness	(2.0–) 2.1 (–2.2) cm
Weight	(9.6–) 9.9 (–10.6) g
Quantity	6
Photographer	D. Fennema

Composition

Triticum	Wheat	grain kernel
–	milk	

Impurities

Avena	Oat	floret

Description

The unpolished grain kernels are medium ground. The shape of the grain fragments is irregular, and the edges of the fragments are sharp. The tarhana is hand-shaped, exhibiting a clear imprint of the finger tips on one face and a curved surface on the opposite face, which is pressed against the palm. An impurity concerns the floret of a wild oat. The Arabic name that is mentioned by the vendor for this tarhana is '*frik paste*'.

FIGURE 305

Tarhana

Obtained	Samsun (Samsun province, Turkey)
Origin	Merzifon district (Amasya province, Turkey)
Date	August 2015
Collector(s)	R. T. J. Cappers & Y. Okur
Context	Shop
GIA-accession	39627
Length/Diameter	(5.1–) 5.9 (–6.7) cm
Thickness	(1.4–) 1.6 (–1.9) cm
Weight	(12.4–) 14.7 (–17.5) g
Quantity	16
Photographer	D. Fennema

Composition

Triticum aestivum ssp. *aestivum*	Bread wheat	grain kernel
–	milk	

Description

The polished grain kernels are coarsely ground. The tarhana units are shaped in the hand, one face shaped by the palm and the other by 3–4 fingers. A similar kind of tarhana was collected in March 2018 in Istanbul at a farmers market, where it was offered for sale by a woman who originates from Merzifon (GIA-41149; not depicted). Apparently, a surplus of this food product is sold as a cash food.

FIGURE 306

Tarhana

Obtained	Tehran (Tehran province, Iran)
Origin	–
Date	April 2017
Collector(s)	R. T. J. Cappers & M. Kaffashi
Context	Shop
GIA-accession	40740
Length/Diameter	7.5 cm
Thickness	1.5 cm
Weight	29.6 g
Quantity	1
Photographer	D. Fennema

Composition

Brassica rapa	Turnip	leaf
Triticum	Wheat	
–	milk	

Impurities

Triticum	Wheat	hull

Description

The unpolished grain kernels are finely ground. The tarhana is mixed with chopped leaves of turnip. It seems to have been shaped into elongated units using two hands, as evidenced by the imprint of the fingertips from one hand and the flat shape of the palm of the other one. The units are pierced in order to thread the tarhana on a thin cord. These units were offered for sale without having been strung.

FIGURE 307

Tarhana

Obtained	Tehran (Tehran province, Iran)
Origin	–
Date	April 2017
Collector(s)	R. T. J. Cappers & M. Kaffashi
Context	Shop
GIA-accession	40739
Length/Diameter	8.6–8.8 cm
Thickness	1.9–2.4 cm
Weight	26.9–33.8 g
Quantity	2
Photographer	D. Fennema

Composition

Brassica rapa	Turnip	leaf
Triticum	Wheat	grain kernel
–	milk	

Description

The unpolished grain kernels are finely ground. The tarhana is mixed with chopped leaves of turnip. It is shaped within a single hand into elongated units showing the imprint of three fingertips and the shape of the palm. The units are pierced in order to thread the tarhana on a thin cord. These units were offered for sale without being strung.

FIGURE 308

Tarhana

Obtained	Tehran (Tehran province, Iran)
Origin	–
Date	April 2017
Collector(s)	R. T. J. Cappers & M. Kaffashi
Context	Shop
GIA-accession	40742
Length/Diameter	8.8–12.8 cm
Thickness	2.3 cm
Weight	30.4–42.2 g
Quantity	2
Photographer	D. Fennema

Composition

Brassica rapa	Turnip	leaf
Triticum	Wheat	grain kernel
–	milk	

Description

The unpolished grain kernels are coarsely ground. The presence of whole grain kernels as well as well as partial kernels suggest that the whole kernels were added after grinding. It seems to be shaped into elongated units using two hands, as evidenced by the imprint of the fingertips from one hand and the flat shape of the palm of the other one. The bottom face has become flattened due to being pressed against the drying surface. The texture is open and the units easily break.

FIGURE 309

Tarhana

Obtained	Tehran (Tehran province, Iran)
Origin	Malayer (Hamedan province, Iran)
Date	April 2017
Collector(s)	R. T. J. Cappers & M. Kaffashi
Context	Shop
GIA-accession	40738
Length/Diameter	11.3–12.5 cm
Thickness	1.0–2.4 cm
Weight	63.2–68.2 g
Quantity	2
Photographer	D. Fennema

Composition

Triticum aestivum ssp. *aestivum*	Bread wheat	bran
Triticum aestivum ssp. *aestivum*	Bread wheat	grain kernel
–	milk	

Description

The unpolished grain kernels are coarsely ground. Some of the grain fragments still have their white endosperm. The duration of par-boiling seems to have differed among the grain kernels that ware used. Small fragments of bran are added. This results in a compact texture with a speckled appearance. The tarhana is shaped into elongated units using two hands, as evidenced by the imprint of the fingertips from one hand and the flat shape of the palm of the other one. The tarhana was made by Luri people from Malayer.

FIGURE 310

Tarhana

Obtained	Tehran (Tehran province, Iran)
Origin	Malayer (Hamedan province, Iran)
Date	April 2017
Collector(s)	R.T.J. Cappers & M. Kaffashi
Context	Shop
GIA-accession	40335
Length/Diameter	(7.3–) 8.0 (–8.9) cm
Thickness	(1.4–) 1.7 (–2.1) cm
Weight	(26.2–) 30.3 (–35.8) g
Quantity	7
Photographer	D. Fennema

Composition

Brassica rapa	Turnip	leaf
Triticum	Wheat	
–	buttermilk	

Description

The unpolished grain kernels are finely ground. The tarhana is mixed with chopped leaves of turnip. Because the smallest fraction is also included, the texture of the final product become solid. The tarhana is shaped into elongated units using two hands, as evidenced by the imprint of 3–4 fingertips from one hand and the flat shape of the palm of the other one. These units were offered for sale strung on a piece of string.

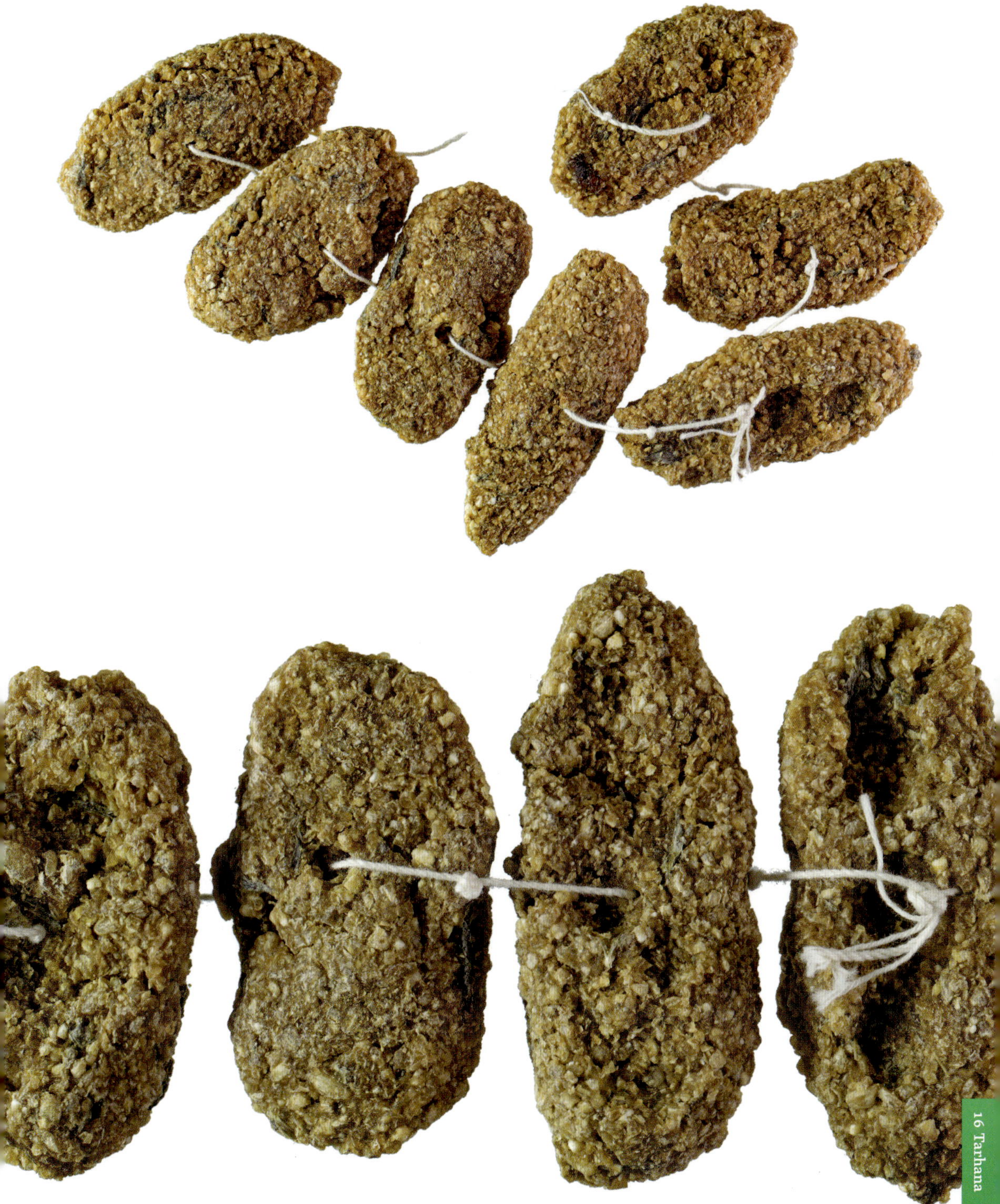

FIGURE 311

Tarhana

Obtained	Baneh (Kurdistan province, Iran)
Origin	–
Date	April 2017
Collector(s)	R. T. J. Cappers, M. Kaffashi & P. Hesami
Context	Shop
GIA-accession	40778
Length/Diameter	(11.3–) 11.9 (–12.3) cm
Thickness	(3.4–) 3.8 (–4.2) cm
Weight	(96.3–) 107.1 (–121.1) g
Quantity	3
Photographer	D. Fennema

Composition

Brassica rapa	Turnip	leaf and stem
Triticum	Wheat	grain kernel
–	milk	

Impurities

Hordeum vulgare ssp. *vulgare*	Hulled 6-row Barley	floret

Description

The unpolished grain kernels are coarsely ground. Many grain kernels have only split lengthwise or are unfragmented, resulting in an open texture. The tarhana is hand-shaped with both hands into thick, elongated units with a deep groove on the upper face. The bottom face is somewhat concave, possibly to reduce the area of contact with the drying surface and to thus speed up the drying process. The shape resembles that of the yellow-coloured tarhana made by Luri people (see the following accessions; pages 596–603).

FIGURE 312

Tarhana

Obtained	Hamedan (Hamedan province, Iran)
Origin	Malayer (Hamedan province, Iran)
Date	July 2017
Collector(s)	R.T.J. Cappers & M. Kaffashi
Context	Bazaar
GIA-accession	40821
Length/Diameter	11.6–12.2 cm
Thickness	3.8 cm
Weight	73.2–73.9 g
Quantity	2
Photographer	D. Fennema

Composition

Curcuma longa	Turmeric	rhizome
Triticum	Wheat	grain kernel
–	milk	

Description

The grain kernels are polished and coarsely ground. The cereal–milk mixture is coloured with turmeric. The tarhana has an open texture. The tarhana is shaped into elongated units using two hands, as evidenced by the imprint of the index finger from one hand and the round shape of the palm of the other one. The greatest thickness is in the middle, and the width gradually narrows towards either end. The shape is reminiscent of the floret of barley, with the imprint of the finger resembling the hilum groove. The tarhana was made by Luri people.

FIGURE 313

Tarhana

Obtained	Tehran (Tehran province, Iran)
Origin	Lorestan province, Iran
Date	April 2017
Collector(s)	R. T. J. Cappers & M. Kaffashi
Context	Bazaar
GIA-accession	40735
Length/Diameter	10.2–10.9 cm
Thickness	3.7–3.9 cm
Weight	64.4–65.3 g
Quantity	2
Photographer	D. Fennema

Composition

Curcuma longa	Turmeric	rhizome
Triticum	Wheat	grain kernel
–	milk	

Description

The grain kernels are polished and coarsely ground. The cereal–milk mixture is coloured with turmeric. The tarhana has an open texture. The tarhana is shaped into elongated units using two hands, as evidenced by the imprint of the index finger from one hand and the round shape of the palm of the other one. The imprint of the finger is somewhat oblique and is deeper on one face. The bottom face shows a faint imprint of a small, square pattern, possibly from curtains. The greatest thickness is in the middle, and the width gradually narrows towards either end. The shape is reminiscent of the floret of barley, with the imprint of the finger resembling the hilum groove. The tarhana was made by Luri people.

FIGURE 314

Tarhana

Obtained	Tehran (Tehran province, Iran)
Origin	Malayer (Hamedan province, Iran)
Date	April 2017
Collector(s)	R.T.J. Cappers & M. Kaffashi
Context	Bazaar
GIA-accession	40737
Length/Diameter	13.6–13.8 cm
Thickness	5.5–5.9 cm
Weight	101.7–103.7 g
Quantity	2
Photographer	D. Fennema

Composition

Curcuma longa	Turmeric	rhizome
Triticum	Wheat	grain kernel
–	milk	

Description

The grain kernels are polished and coarsely ground. The cereal–milk mixture is coloured with turmeric. The tarhana is shaped into large, elongated units by using a mould. The location and shape of the small groove differs among the two specimens, indicating that they were produced in different moulds. The diamond-shaped impression in the bottom face differs in orientation between the two specimens and must therefore have been produced by the surface against which the tarhana units were pushed and on which they were subsequently dried, rather than by the mould. The greatest thickness is in the middle, and the width gradually narrows towards either end. The shape is reminiscent of the floret of barley, with the imprint resembling the hilum groove. The tarhana has an open texture and is rather fragile. When the tarhana was collected, it had a bright yellow colour, which is still visible in photos taken shortly after. After a couple of months, the colour had partly faded away, and the colour is thus indicative of the product's age. The tarhana was made by Luri people from Malayer.

FIGURE 315

Tarhana

Obtained	Hamedan (Hamedan province, Iran)
Origin	Malayer (Hamedan province, Iran)
Date	July 2017
Collector(s)	R. T. J. Cappers & M. Kaffashi
Context	Bazaar
GIA-accession	40822
Length/Diameter	16.4–17.9 cm
Thickness	0.9–5.1 cm
Weight	158.2–161.1 g
Quantity	2
Photographer	D. Fennema

Composition

Curcuma longa	Turmeric	rhizome
Triticum	Wheat	grain kernel
–	milk	

Impurities		
–	stone	

Description

The grain kernels are polished and coarsely ground. The cereal–milk mixture is coloured with turmeric. The tarhana is shaped into large, elongated units by using a mould. The greatest thickness is in the middle, and the width gradually narrows towards either end. The shape is reminiscent of the floret of barley, with the imprint resembling the hilum groove. The tarhana has an open texture and is rather fragile. When the tarhana was collected, it had a bright yellow colour, which is still visible in the photos. After a couple of months, the colour partly faded away, and the colour is thus indicative of the product's age. The tarhana was made by Luri people.

FIGURE 316

Tarhana

Obtained	Nigde (Nigde province, Turkey)
Origin	–
Date	August 2014
Collector(s)	F. Fantone
Context	–
GIA-accession	40972
Length/Diameter	9.2–9.8 cm
Thickness	4.4–5.7 cm
Weight	83.2–85.8 g
Quantity	2
Photographer	D. Fennema

Composition

Triticum aestivum ssp. *aestivum*	Bread wheat	grain kernel
–	milk	

Description

The unpolished grain kernels are coarsely ground. The tarhana is hand-shaped into convex-based, pyramidal units.

FIGURE 317

Captions to the full-page photos

A number of photos are presented in full-page format as an introduction to a chapter or other part of the book. The captions to these photos are given below. For all other photos, the captions are presented below the photos.

PAGE 2 Tarhana made from wheat and milk and shaped in the form of a barley floret. This food item has a long storage life and symbolizes the integration of early crop husbandry and animal husbandry in south-west Asia. It was collected in Hamedan (Iran) in July 2017 (DF).

PAGE 8 Transporting a load of barley (*Hordeum vulgare*) from the field to the road. The barley is harvested with a sickle (Kalamafka, Crete; July 2016; RC).

PAGE 12 A large flock of sheep and goats, as well as a donkey, visit a cereal field after the harvest. The short arable weeds that are eaten by these animals will eventually become available to humans as wool, milk, and meat. Such a large number of sheep and goats will produce a surplus of milk that has to be processed into a long-life product daily. In the foreground, some turkeys are visible (Boskin, Iraq; June 2014; RC).

PAGE 43 A woman demonstrates the polishing of naked wheat (*Triticum*). The grain kernels are placed in a stone mortar with a conical cavity. The grain is wetted with some water and the pounding is done with a stone pestle that weighs 8 kg (Abu Bakra, Iraq; June 2014; RC).

PAGE 75 A quantity of kashk offered for sale in a bazaar. The fracture surfaces and the variation in size, shape, and colour suggest that we are dealing with a mixture of remnants originating from different supplies (Tabriz, Iran; July 2017; RC).

PAGE 92 An agadir built at the top of a mountain. The agadir accommodates small rooms for each of the local farmers to store the harvest of their crops. Each room is protected with a reinforced door, and the agadir is continually monitored by a guard (Amtoudi, Morocco; April 2007; RC).

PAGE 99 Specially shaped loaves of dough that are placed in a peel oven for a single baking, together with loaves of dough that will be baked twice (north-west of Psathi, Crete, Greece; July 2016; RC).

PAGE 105 A wall painting in the tomb of Meryneith, in the New Kingdom necropolis at Saqqara, showing the cleaning by hand picking of what appear to be grains or seeds spread out on a tray. Such cleaning can be done prior to storage or prior to food preparation (Giza Governorate, Egypt; February 2010; RC).

PAGE 106 An assemblage of used domestic metalware offered for sale in a bazaar. It includes kettles (*kazans*) for wet-heating, a tray that can be used for the proving of shaped dough, and ewers of Kurdish origin (Sanandaj, Iran; April 2017; RC).

PAGE 128 Sprouted Bread wheat (*Triticum aestivum* ssp. *aestivum*) offered for sale at a bazaar. The sprouted grain is used for celebrating the Iranian New Year. Other plants that are used include Maize (*Zea mays*), Lentil (*Lens culinaris*), and Mung bean (*Vigna radiata*). Sometimes, a knot is made in some of the plants (Tehran, Iran; April 2017; RC).

PAGE 132 Roasted grain of Bread wheat (*Triticum aestivum* ssp. *aestivum*) mixed with roasted fruits of Turpentine tree (*Pistacia terebinthus*) and seeds of what is probably Apricot (cf. *Prunus armeniaca*). The mixture is home-made in Sanadaj and is stored in a recycled glass jar. Judging from the presence of scorch marks, the grain was roasted without having been covered by sand (Tehran, Iran; July 2017; RC).

PAGE 166 Spikes of Hard wheat (*Triticum turgidum* ssp. *durum*) are raked up and set on fire. The isolated spikes in the foreground show that only a small part of the culm remains attached to the spike. Therefore, additional fuel is needed for roasting the grain (Ajaa, Palestine; May 2016; HM).

PAGE 202 A flat bread being baked on a convex metal plate. The metal plate is fixed onto a metal ring. Wood from different plants is used as fuel. The metal tong in front is used for adjusting the fuel. This metal oven sits in a small room that is specially designed for food preparation. A chimney is connected to the oven in order to prevent smoke from entering the room (Novin, Iran; April 2017; RC).

PAGE 268 Bulgur made of Emmer (*Triticum turgidum* ssp. *durum*) offered for sale at a bazaar. Placed in front of the shop, such supplies are an attractive food supply for birds. This bulgur will require further cleaning by hand picking. Impurities in the sample we obtained include rachillate spikelets of the wheat. The bulgur is rinsed before cooking to remove any bird droppings (Baneh, Iran; April 2017; RC).

PAGE 340 Women making home-made pasta. The dough is first shaped into long cylinders and then cut into small pieces with a knife. Each piece is first flattened and then rolled up with the fingertips (Kalamafka, Crete, Greece; July 2017; RC).

PAGE 372 A package of couscous. The couscous is home-made by Palestinian women (obtained in Utrecht, the Netherlands; February 2017; DF).

PAGE 394 A home-made kashk (left) and an industrially produced kashk (right) offered for sale in a shop (Kandovan, Iran; July 2017; RC).

PAGE 450 Different qualities of qara qurut offered for sale in a bazaar. The small units of white gum, wrapped in plastic, are from the Turpentine tree (*Pistacia terebinthus*) (Shiraz, Iran; July 2017; RC).

PAGE 472 A woman crumbling tarhana into irregular units on the roof of her house. The tarhana is spread out on sheets of plastic and then sun-dried (Kalamafka, Crete, Greece; July 2016; RC).

PAGE 606 Slicing of the barley bread starts immediately after the first baking so that the slices can quickly be put back into the oven before the oven starts losing too much heat. In order to facilitate this slicing (breaking), incisions are made in the dough before the first baking (west of Katharon, Crete; July 2016; RC).

PAGE 610 Wall painting in the tomb of Meryneith (Saqqara, New Kingdom necropolis) showing solid and liquid foods (Giza Governorate, Egypt; February 2010; RC).

PAGE 618 A label on a large chest of drawers mentioning barley from India. This kind of plant material is considered to be of special value, and only a limited quantity is offered for sale in this drugstore located in Khan al-Khalili (Cairo, Egypt; March 2001; RC).

PAGE 628 Collecting vegetables and aromatic plants on a hillside. The woman uses a tuck-pointing trowel and a knife to uproot the plants she is looking for. The plants are stored in a cloth that is tied behind her back (Quellace, Iran; April 2017; RC).

References

Abbas A., N. Tabanca, G. Ozek, T. Ozek, Z. Aytac, U. R. Bernier, N. M. Agramonte, K. H. C. Baser & I. A. Khan (2015): Essential oils of *Echinophora lamondiana* (Apiales: Umbelliferae): A relationship between chemical profile and biting deterrence and larvicidal activity against mosquitoes (Diptera: Culicidae). *Journal of Medicinal Entomology* 52 (1), 93–100.

Abdelrahman, R. A., A. M. Adel & I. Smetanska (2010): *Fermented milk products*. New York: Nova Science Publishers.

Abou-Donia, S. A. (1996): Manufacture of Egyptian, soft and pickled cheeses. In: R. K. Robinson & A. Y. Tamime (Eds.): *Feta and related cheeses*. Cambridge: Woodhead Publishing, pp. 166–208.

Ahmad, S. A. & A. A. Askari (2015): Ethnobotany of the Hawraman region of Kurdistan Iraq. *Harvard Papers in Botany* 20 (1), 85–89.

Al-Azm, A. N. (2009): The importance and antiquity of frikkeh: A simple snack or a socio-economic indicator of decline and prosperity in the ancient Near East. In: A. S. Fairbairn & E. Weiss (Eds.): *From foragers to farmers: Papers in honour of Gordon C. Hillman*. Oxford: Oxbow Books, pp. 112–116.

Amiryousefi, M. R. & S. M. A. Razavi (2013): Viscous flow behavior, microstructure and physiochemical characteristics of a sweet paste functional dessert (samanu) made from germinated wheat. *Journal of Food Process Engineering* 36, 385–396.

Anderson, P. C. (1999): Experimental cultivation, harvest, and threshing of wild cereals: Their relevance for interpreting the use of Epipaleolithic and Neolithic artefacts. In: P. C. Anderson (Ed.): *Prehistory of agriculture: New experimental and ethnographic approaches*. Cotsen Monograph 40. Los Angeles: Institute of Archaeology, UCLA, pp. 118–144.

Anderson, P. C. & F. Sigaut (2014): Introduction: Reasons for variability in harvesting techniques and tools. In: P. C. Anderson & L. Peňa-Chocarro (Eds.): *Early agricultural remnants and technical heritage (EARTH): 8,000 years of resilience and innovation*. Vol. 2. Oxford: Oxbow Books, pp. 85–92.

Asouti, E. & D. Q. Fuller (2013): A contextual approach to the emergence of agriculture in south-west Asia: Reconstructing early Neolithic plant-food production. *Current Antropology* 34, 299–345.

Aubaile-Sallenave, F. (1994): 'Al-Kishk'. In: S. Zubaida & R. Tapper (Eds.): *Culinary cultures of the Middle East*. London: I. B. Tauris Publishers, pp. 105–139.

Avitsur, S. (1975): The way to bread: The example of the land of Israel. *Tools and Tillage* 2 (4), 228–241.

Baik, B. K. (2014): Processing of barley for food and feed. In: P. R. Shewry and S. E. Ullrich (Eds.), *Barley: Chemistry and technology* (2nd ed.). St. Paul: AACC International, pp. 233–268.

Behre, K.-E. (1999): The history of beer additives in Europe—A review. *Vegetation History and Archaeobotany* 8, 35–48.

Briggs, D. E. (1978): *Barley*. London: Chapman & Hall.

Cantrell, P.A. (2000): Beer and ale. In: K.F. Kiple & K.C. Ornelas (Eds.): *The Cambridge world history of food*. Cambridge: Cambridge University Press, pp. 619–625.
Cappers, R.T.J. & R. Neef (2012): *Handbook of plant palaeoecology*. Groningen Archaeological Studies 19. Groningen: Barkhuis and Groningen University Library.
Cappers, R.T.J., R. Neef, R.M. Bekker, F. Fantone & Y. Okur (2016): *Digital atlas of traditional agricultural practices and food processing*. Groningen Archaeological Studies 30. Groningen: Barkhuis and Groningen University Library. [http://www.plantatlas.eu].
Carretero, L.G., M. Wollstonecroft & D. Fuller (2017): A methodological approach to the study of archaeological cereal meals: A case study at Çatalhöyük East (Turkey). *Vegetation History and Archaeobotany* 26 (4), 415–432.

Dagher, S.M. (1991): Traditional foods in the Near East. FAO *Food and Nutrition* Paper 50. Rome: FAO.
Değirmencioğlu, N., D. Göçmen, A. Dağdelen & F. Dağdelen (2005): Influence of Tarhana herb (*Echinophora sibthorpiana*) on fermentation of tarhana, Turkish traditional fermented food. *Food Technology and Biotechnology* 43 (2), 175–179.
Delcour, J.A. & R.C. Hoseney (2013): *Principles of cereal science and technology* (3rd ed.). St. Paul: AACC International.
Dunham, C. (1987): *The hidden Himalayas*. New York: Abbeville Press Publishers.

Edelberg, L. & S. Jones (1979): *Nuristan*. Graz: Akademische Druck-u. Verlagsanstalt.
Eitam, D., M. Kislev, A. Karty & O. Bar-Yosef (2015): Experimental barley flour production in 12,500-year-old rock-cut mortars in south-western Asia. *Plos One* 10 (7) e0133306. doi:10.1371/journal.pone.0133306.
Ekvall, R.B. (1968): *Fields on the hoof: Nexus of Tibetan nomadic pastoralism*. New York: Holt, Rinehart and Winston.
El-Gendy, S.M. (1983): Fermented foods in Egypt and the Middle East. *Journal of Food Protection* 46 (4), 358–367.
El-Haddad, L. & M. Schmitt (2011): Gaza's food heritage. *Saudi Aramco World* 62 (6), 2–11.
Elhamid, A.M.A. (2016): Physicohemical, rheological and sensory properties of Egyptian kariesh cheese containing wheat bran. *International Journal of Dairy Technology* 69 (3), 425–432.
Erkal, H., S. Çelik, B. Bilgi & H. Köksel (2006): A new approach for the utilization of barley in food products: Barley tarhana. *Food Chemistry* 97, 12–18.
Ertuğ, Z.F. (2009): Wild plant foods: Routine dietary supplements or famine foods? In: A. Fairbairn & E. Weiss (Eds.): *From foragers to farmers: Papers in honour of Gordon C. Hillman*. Oxford: Oxbow Books, pp. 64–70.
Evershed R.P., S. Payne, A.G. Sherratt, M.S. Copley, J. Coolidge, D. Urem-Kotsu, K. Kotsakis, M. Özdoğan, A.E. Özdoğan, O. Nieuwenhuyse, P.M.M.G. Akkermans, D. Bailey, R.-R. Andeescu, S. Campbell, S. Farid, I. Hodder, N. Yalman, M. Özbaşaran, E. Bıçakcı, Y. Garfinkel, T. Levy & M.M. Burton (2008): Earliest date for milk use in the Near East and southeastern Europe linked to cattle herding. *Nature* 455, 528–531.

Farahani, G., H. Ezzatpanah & S. Abbasi (2014): Characterization of siahmazgi cheese, an Iranian ewe's milk variety: Assessment of physico-chemical, textural and rheological specifications during ripening. *Food Science and Technology* 58, 335–342.
Ferdinand, K. (1993): *Bedouins of Qatar*. London: Thames and Hudson.
Frederiksen, B. (1996): *Caravans and trade in Afghanistan: The changing life of the nomadic Hazarbuz*. London: Thames and Hudson.
Fuller, D.Q., R.G. Allaby & C. Stevens (2010): Domestication as innovation: The entanglement of techniques, technology and chance in the domestication of cereal crops. *World Archaeology* 42, 13–28.

Garthwaite, G.R. (2009): *Khans and Shahs: A history of the Bakhtiyari tribe in Iran*. London: I.B. Tauris Publishers.
Goldstein, M.C. & C.M. Beall (1990): *Nomads of western Tibet: The survival of a way of life*. London: Serindia Publications.
Goudsblom, J. (1992): *Vuur en beschaving*. Amsterdam: Meulenhoff.
Grammet, I. (1998): De graanopslagplaatsen. In: I. Grammet & M. de Meersman (Eds.): *Magisch Marokko*. Tervuuren: Koninklijk Museum voor Midden-Afrika, pp. 62–65.

Halstead, P. (2014): *Two oxen ahead: Pre-mechanized farming in the Mediterranean*. Chichester: Wiley Blackwell.
Hamilton, R.W. (2003): Labor, ritual, and the cycle of time. In: R.W. Hamilton (Ed.): *The art of rice: Spirit and sustenance in Asia*. Los Angeles: Fowler Museum of Cultural History, UCLA, pp. 37–61.
Hammer, K. (1984): Das Domestikationssyndrom. *Kulturpflanze* 32, 11–34.
Harlan, J.R. (1992): *Crops and man*. Madison: American Society of Agronomy.
Heiss, A.G., F. Antolin, N. Bleicher, C. Harb, S. Jacomet, M. Kühn, E. Marinova, H.-P. Stika & S.M. Valamoti (2017): State of the (t)art: Analytical approaches in the investigation of components and production traits of archaeological bread-like objects, applied to two finds from the Neolithic lakeshore settlement Parkhaus Opéra (Zürich, Switzerland). *Plos One* 12 (8): e0182401.
Herken, E.N. & A.H. Çon (2014): Use of different lactic starter cultures in the production of tarhana. *Journal of food processing and preservation* 38, 59–67.
Hermanns, M. (1949): *Die Nomaden von Tibet: Die sozial-wirtschaftlichen Grundlagen der Hirtenkulturen in A mdo und von Innerasien: Ursprung und Entwicklung der Viehzucht*. Vienna: Verlag Herold.
Hilali, M., E. El-Mayda & B. Rischkowsky (2011): Characteristics and utilization of sheep and goat milk in the Middle East. *Small Ruminant Research* 101, 92–101.
Hildalgo, A. & A. Brandolini (2013): Nutritional properties of Einkorn wheat (*Triticum monococcum* L.). *Journal of the Science of Food and Agriculture* 94, 601–612.
Hillman, G.C. & M.S. Davies (1992): Domestication rate in wild wheats and barley under primitive cultivation: Preliminary results and archaeological implications of field measurements of selection coefficient. In: P.C. Anderson (Ed.): *Préhistoire de l'agriculture: nouvelles approches expérimentales et ethnographiques*. CRA Monografie 6. Paris: CNRS, pp. 113–158.
Hobbs, J.J. (1989): *Bedouin life in the Egyptian wilderness*. Cairo: American University in Cairo Press.
Huang, J. (2009): *Tribeswomen of Iran: Weaving memories among Qushqa'i nomads*. London: I.B. Tauris Publishers.

Jabbur, J. S. (1995): *The Bedouins and the desert: Aspects of nomadic life in the Arab East*. New York: State University of New York Press.
Jones, S. (1996): *Tibetan nomads: Environment, pastoral economy, and material culture*. London: Thames and Hudson.

Karaoğlu, M. M. & H. G. Kotancilar (2006): Kavut, a traditional Turkish cereal product: Production method and some chemical and sensorial properties. *International Journal of Food Science and Technology* 41, 233–241.
Kivanç, M. & E. G. Funda (2017): A functional food: A traditional tarhana fermentation. *Food Science and Technology* 37 (2), 269–274.
Köksel, H., M. J. Edney & B. Özkaya (1999): Barley bulgur: Effect of processing and cooking on chemical composition. *Journal of Cereal Science* 29, 185–190.
Kraybill, N. (1977): Pre-agricultural tools for the preparation of foods in the Old World. In: C. A. Reed (Ed.): *Origins of agriculture*. The Hague, Mouton Publishers, pp. 485–522.

Leach, H. M. (1990): Food processing technology: its role in inhibiting or promoting change in staple foods. In: C. Gosden & J. Hather (Eds.): *The prehistory of food. Appetites for change*. London: Routledge.

Mauger, T. & D. Mauger (1986): *A l'ombre des tentens noires*. Riyad: Tihama.
McGee, H. (2004): *On food and cooking: The science and lore of the kitchen.* New York: Scribner.
Meurers-Balke, J. & J. Lüning (1990): Experimente zur Verarbeitung von Spelzgetreiden. In: F. Fansa (Ed.): *Experimentelle Archäologie in Deutschland.* Archäologische Mitteilungen aus Nordwestdeutschland. Beihefte 4. Oldenburg: Isensee Verlag, pp. 93–112.
Mortensen, I. D. (1993): *Nomads of Luristan: History, material culture, and pastoralism in western Iran*. London: Thames and Hudson.
Musselman, L. J. & A. B. Al-Mouslem (2001): *Triticum durum* in northern Syria: Parched corn (frikeh) of the Bible? *Economic Botany* 55 (2), 187–189.

Nederlands Voedingsstoffenbestand (NEVO). Online version 2016/5.0, RIVM, Bilthoven.
Nicolaisen, J. & I. Nicolaisen (1997): *The pastoral Tuareg: Ecology, culture, and society*. London: Thames and Hudson.

Oudenhoven, F. van & J. Haider (2015): *With our own hands: A celebration of food and life in the Pamir mountains of Afghanistan and Tajikistan.* Utrecht: Stichting LM Publishers.
Özçam, M., E. Obuz & H. Tosun (2014): Aflatoxin M1 in tarhana chips. *Food Additives & Contaminants*: Part B 7 (3), 182–185.

Palmer, C. (2002): Milk and cereals: Identifying food and identity among Fallāhīn and Bedouin in Jordan. *Levant* 34, 173–195.
Pedersen, G. (1994): *Afghan nomads in transition: A century of change among the Zala Khān Khêl*. London: Thames and Hudson.
Pederson, P. N., T. Richter & A. Arranz-Otaegui (2016): Preliminary analysis of the Late Natufian ground stone from Shubayqa 1, *Jordan. Journal of Lithic Studies* 3 (3), 379–402.

Peña-Chocarro, L. (2014): The use of mesorias to harvest hulled wheat by stripping: An ancient tool? In: P. C. Anderson & L. Peña-Chocarro (Eds.): *Early agricultural remnants and technical heritage (EARTH): 8,000 years of resilience and innovation*. Vol. 2. Oxford: Oxbow Books, pp. 103–105.

Peña-Chocarro, L. & L. Zapata (2014): Parching and dehusking hulled wheats. In: P. C. Anderson & L. Peña-Chocarro (Eds.): *Early agricultural remnants and technical heritage (EARTH): 8,000 years of resilience and innovation*. Vol. 2. Oxford: Oxbow Books, pp. 226–232.

Peña-Chocarro, L. L. Zapata Peña, J. E., González Urquijo & J. J. Ibáñez Estévez (2009): Einkorn (*Triticum Monococcum* L.) cultivation in mountain communities of the western Rif (Morocco): An ethnoarchaeological project. In: A. Fairbairn & E. Weiss (Eds.), *From foragers to farmers: Papers in honour of Gordon C. Hillman*. Oxford: Oxbow Books, pp. 103–111.

Percival, J. (1921): *The wheat plant*. London: Duckworth.

Procopiou, H. (2014): Barley meal processing in the Aegean world: A look at diversity. In: P. C. Anderson & L. Peña-Chocarro (Eds.): *Early agricultural remnants and technical heritage (EARTH): 8,000 years of resilience and innovation*. Vol. 2. Oxford: Oxbow Books, pp. 243–246.

Qarooni, J. (1996): *Flat bread technology*. New York: Chapman & Hall.

Samuel, D. (2000): Brewing and baking. In P. T. Nicholson & I. Shaw (Eds.): *Ancient Egyptian materials and technology*. Cambridge: Cambridge University Press, pp. 537–576.

Shaker, R. R., R. Y. Jumah, B. Tashtoush & A. F. Zraiy (1999): Manufacture of jameed using a spray drying process. *International Journal of Dairy Technology* 52 (3), 77–80.

Shoemaker, A. C., M. I. J. Davies & H. L. Moore (2017): Back to the grindstone? The archaeological potential of grinding-stone studies in Africa with reference to contemporary grinding practices in Marakwet, northwest Kenya. *African Archaeological Review* 34, 415–435.

Spjut, R. W. (1994): *A systematic treatment of fruit types*. Memoirs of the New York Botanical Garden 70. New York: New York Botanical Garden Printing Department.

Stanier, R. Y., E. A. Adelberg & J. L. Ingraham (1981): *General microbiology* (4th ed.). London: Macmillan.

Steinkraus, K. H. (1983): Lactic acid fermentation in the production of foods from vegetables, cereals and legumes. *Antonie van Leeuwenhoek* 49, 337–348.

Tamer, C. E., A. Kumral, M. Aşan & I. Şahin (2007): Chemical compositions of traditional tarhana having different formulations. *Journal of Food Processing and Preservation* 31, 116–126.

Tamime, A. Y., M. Hickey & D. D. Muir (2014): Stained fermented milks: A review of existing legislative provisions, survey of nutritional labelling of commercial products in selected markets and terminology of products in some selected countries. *International Journal of Dairy Technology* 67, 305–333.

Tamime, A. Y. & T. P. O'Connor (1995): Kishk: A dried fermented milk/cereal mixture. *International Dairy Journal* 5, 109–128.

Todaro, A., F. A. Adly & O. A. H. Omar (2013): History, processing and quality enhancement of traditional Egyptian kariesh cheese: A review. *Food Science and Technology* 1 (1), 1–6.

Tzarfati, R., Y. Saranga, V. Barak, A. Gopher, A. B. Korol & S. Abbo (2013): Threshing efficiency as an incentive for rapid domestication of emmer wheat. *Annals of Botany* 112, 829–837.

Valamoti, S. M. (2011): Ground cereal food preparations from Greece: The prehistory and modern survival of traditional Mediterranean 'fast foods'. *Archaeological and Anthropological Sciences* 3, 19–39.

Veen, M. van der (1989): Charred grain assemblages from Roman-period corn driers in Britain. *Archaeological Journal* 146, 302–319.

Vigne, J.-D. (2015): Early domestication and farming: What should we know or do for a better understanding? *Anthropozoologica* 50, 123–150.

Watson, P.J. (1979): *Archaeological ethnography in western Iran*. Viking Fund Publications in Anthropology 57. Arizona: University of Arizona Press.

Weir, S. (1990): *The bedouin*. London: British Museum Publications.

Willcox, G. (2005): The distribution, natural habitats and availability of wild cereals in relation to their domestication in the Near East: Multiple events, multiple centres. *Vegetation History and Archaeobotany* 14, 534–541.

Wilson, R. T. (2017): Traditional milk processing and value-added dairy products in selected Arab countries. *International Journal of Dairy Technology* 70 (3), 307–319.

Zeder, M. A. (2011): The origins of agriculture in the Near East. *Current Anthropology* 52 (4), 221–235.

Zeist, W. van & S. Bottema (1991): Late Quaternary vegetation of the Near East. In: *Beihefte zum Tübinger Atlas des Vorderen Orients: Reihe A (Naturwissenschaften)* 18, 1–156.

هندى شعير
INDIAN BARLEI

Glossary

Acid-coagulated cheese Alternative term: *'non-rennet cheese'*.

Alcoholic fermentation Metabolic pathway that releases stored energy by converting organic compounds (such as sugar) into compounds with less stored energy, including alcohol (ethanol).

Aleuron(e) layer Outer layer of the endosperm that is rich in proteins and enzymes. The number of cell layers varies among cereals; it consists of a single cell layer in Wheat (*Triticum*) and of three cell layers in Barley (*Hordeum*).

Baking Cooking technique that makes use of dry-heating in which heat transfer by radiation and hot-air convection predominates.

Batter Liquid mixture of meal and/or flour, water, and salt. If florets of barley are pulverized, the batter also contains small fragments of hulls. Usually, sourdough or yeast is also added.

Beer An alcoholic beverage made from cereal grain and water. The alcohol is obtained by enzymatic conversion of starch into sugars during germination of the grain and subsequent alcoholic fermentation of sugars by yeast.

Black cheese Alternative term: *'qara qurut'*.

Boiling Wet-heating in water at a temperature of 100 °C.

Bran The outer part of the grain kernel, consisting of the fruit wall, the seedcoat, and the aleurone layer.

Bread Baked unleavened or leavened loaves made from batter or dough.

Bulgur Dried par-boiled and granulated grain. The polishing of the grain before granulation is optional.

Butter Solid milk product consisting of concentrated fat. The milk fat is concentrated by first churning milk and then separating out and removing the buttermilk. The word butter may refer to either ordinary butter or clarified butter.

Buttermilk Liquid, de-fatted by-product of butter making.

By-product In cereal processing: post-harvesting product of secondary interest produced at different stages of fragmentation (in cereals: threshing, de-floretting, de-graining, pearling, and polishing) and subsequent separation. In cereal processing there are two kinds of by-product, (1) the vegetative plant parts if grain is the principal product or (2) the grain if culms are the principal product. In milk processing: whey. Alternative term: *'secondary product'*.

Caryopsis Dry fruit with fruit wall (pericarp) and seedcoat (testa) closely adhering and fused in the hilar region (*sensu* Spjut, 1994). Typical for the Grass family (Poaceae).

Cereal-milk product Food product in which processed cereal grains and processed milk are mixed. See also: tarhana.

Chaff Mixture of the vegetative parts of the infructescence that is separated from the harvest by threshing, de-floretting, and de-graining. The vegetative parts may include (fragments) of the glumes, lemmas, and paleas (possibly present as sterile florets); awns (from glumes and/or lemmas); and rachis fragments.

Chakka Non-rennet cheese made from buttermilk. The buttermilk is separated from fermented milk and concentrated by removing the whey.

Cheese Solid milk product made by coagulation of proteins and removal of whey. A large proportion of the milk fat may have been removed by butter making prior to the coagulation of proteins. See also: non-rennet cheese and rennet cheese.

Churning Of milk: stirring (mixing by using a tool [stirrer] that moves in the liquid) or shaking (mixing by moving the container) to solidify the fat.

Clarified butter Butter from which most of the carbohydrates and proteins have been removed to improve the shelf-life. Clarified butter can be solid or liquid. See also: ghee.

Cooking Food preparation by the application of heat.

Couscous In the strict sense: dried intermediate food consisting of granules of a uniform size. It is made by mixing coarse and fine granulated grain with water. Salt can be added. In its basic form it is made without heating. In the broad sense: any dish made with couscous.

Cream Solid milk fat obtained by skimming. Once the cream is skimmed off, the remaining milk is called skimmed milk.

Cream cheese In this publication: non-rennet cheese made from skimmed milk, which is obtained from fermented milk. The term is also used elsewhere for dehydrated cream, which is considered here to be a kind of butter. The type-name for cream cheese in this publication is '*kariesh*'.

Creaming Separating raw milk into cream and skimmed milk, resulting from a gravity-based de-fatting process.

Cuisine Style of cooking that is characterized by a combination of dishes.

Curd In the strict sense: solidified casein proteins. In the broad sense: the solid part of milk, consisting of a mixture of solidified casein proteins and water with suspended and dissolved nutrients.

Curdling In milk: solidifying due to the coagulation of the casein proteins. The coagulation can be caused by a lowering of the acidity (caused by fermentation of lactic acid bacteria) and/or by enzyme activity (see: rennin and rennet).

Dairy product Alternative term: '*milk product*'.

De-awning Fragmentation process by which awns are removed.

De-branning Fragmentation process by which the bran becomes isolated from the starchy endosperm. Note that this term is not a synonym of polishing because the bran does not include the germ.

De-fatted milk product Liquid milk product obtained by the removal of clumped fat. Two kinds are recognized: (1) skimmed milk (by removal of cream) and (2) buttermilk (by removal of the butter).

De-floretting Fragmentation of the rachillate triplet. Fragmentation is followed by separation. The principal product is a floret.

De-germing Fragmentation process by which the germ becomes entirely or partially separated from the grain kernel. The principal product is a grain kernel without germ.

De-graining Fragmentation of the rachillate spikelet. Fragmentation is followed by separation. The principal product is a grain kernel. Preferred term for *'de-hulling'*.

De-hulling In barley and wheat: alternative but not preferred term for *'de-graining'*. Alternative terms of de-hulling: *'de-husking'*, *'hulling'*, and *'husking'*.

De-husking Preferred term: *'de-graining'*.

Diaspore Dissemination or dispersal unit of a plant. It can consist of (1) a seed, (2) a seed surrounded by (part of) the fruit or (3) a seed surrounded by (part of) the fruit and an exocarp.

Dish Separate unit of a meal that is made from one or more ingredients.

Dough Solid mixture of meal and/or flour, water and salt. If florets of barley are pulverized, the dough also contains small fragments of hulls. Usually, sourdough or yeast is also added.

Dried milk product Incompressible milk product. See also: tarhana and solid milk product.

Drying Reducing the moisture content without the chemical transformation caused by exposure to heat.

Embryo Young plant present in a seed. See also: germ.

Endosperm The interior of a seed, surrounded by the seedcoat. It is divided into two parts: (1) the outer part, which is the aleurone layer, and (2) the inner part, which is storage tissue. In cereals, the endosperm is well developed and the inner part is rich in starch and, depending on the genus, also in gluten.

Fermentation Metabolic pathway that releases stored energy by converting organic compounds (such as sugar) into compounds with less stored energy.

Floret Unit within a spikelet. At the flowering stage: lemma, palea, and the enclosed flower. At the fruiting stage: lemma, palea, and the enclosed grain kernel. See also: spikelet.

Flour Pulverized polished grain, thus with bran and germ removed. See also: meal.

Free-threshing (in cereals) We advocate avoiding the term free-threshing because it is confusing. Rachis brittleness (the difference in rachis toughness) and hull tightness (the extent to which the lemma and palea are connected with the grain kernel) are linked features with respect to dispersal and artificial fragmentation. In domesticated wheat, there are two kinds of rachis brittleness, namely, semi-brittle and non-brittle. Only wheat taxa with a non-brittle rachis are designated as free-threshing taxa (the principal product of threshing is a grain kernel). In domesticated barley, there is only one kind of rachis brittleness, namely, non-brittle. Therefore all domesticated barley is categorized as *'free-threshing'*, irrespective of its hull tightness (the principal products of threshing are either florets or grain kernels). In domesticated barley, the term free-threshing does not express the difference in hull tightness. Preferred term: *'naked'* (for a detailed discussion, see Cappers et al., 2016, pp. 1311–1318).

Fresh milk Alternative term: *'raw milk'*.

Frikeh Roasted unripe grain kernels. The grain kernels are usually coarsely ground.

Fruit Ripe ovary, containing one or more seeds.

Fruit wall The outer structure of a mature fruit that develops from the ovary wall after fertilization. The fruit wall may develop into different layers. The term fruit wall is specifically used in fruit types in which fruit and seed are fused, such as in the caryopsis. Latin term: pericarp.

Frying Heating by using oil or fat.

Galiyya Roasted ripe grain kernels.

Germ Mill fraction of the seed or 1-seeded fruit (caryopsis), concerning the embryo in the seed. In grain kernels, it is located at the dorsal side, at the base, and is only covered by the seedcoat and fruit wall.

Germination The onset of growth of a seed.

Ghee Liquid butter clarified by heating. The liquid fat is separated from any remaining curds and from water with dissolved and soluble nutrients.

Glume Empty bract associated with one or more florets.

Glume wheat We advocate avoiding the term glume wheat because it is confusing. The threshing unit of a so-called glume wheat also contains other chaff particles, and the grain kernel is primarily covered by the lemma and palea. Moreover, two glumes are present in each spikelet of all wheat taxa, irrespective of the hull tightness. Preferred term: '*hulled wheat*'.

Grain kernel Fruit of a cultivated or domesticated grass.

Granulation Fragmentation into coarse fragments.

Grinding Fragmenting involving a horizontal movement that is usually done with grinding stones. Alternative but not preferred term when referring to grinding by means of a mill: 'milling'.

Harvesting Collecting whole plants or separating plant parts that contain the principal product from the undesired parts. This process is mostly confined to plants that are grown in gardens, fields, or orchards.

Harvesting knife Harvesting implement used for cutting crops. It consists of a short, straight blade that is fitted to a short handle.

Harvesting unit Single specimen of the harvest. It consists of a whole plant or plant part. In barley and wheat, three kinds of harvesting units are recognized: (1) rachillate triplet (wild barley); (2) rachillate spikelet (wild wheat); (3) culm and spike (domesticated barley and domesticated wheat).

Hull(s) Collective noun for glumes, lemmas, and paleas. Alternative term: '*husk(s)*'.

Hulled (in cereals) Cereal in which the grain kernels are tightly enclosed by their hulls (lemma; palea; and, depending on the genus, also the glumes). The harvesting and/or threshing unit is a rachillate triplet, a rachillate spikelet, or a floret. See also: glume wheat.

Hulling Preferred term: '*de-graining*'.

Hull tightness The extent to which lemma and palea are connected with the grain kernel.

Husking Preferred term: '*de-graining*'.

Inflorescence The way in which flowers are grouped.

Infructescence The way in which fruits are grouped.

Ingredient Vegetal or animal product that is part of a food product.

Intermediate food Processed food with a long shelf-life that needs to be further processed prior to consumption.

Kariesh Non-rennet cheese made from skimmed milk. The skimmed milk is separated from fermented milk and concentrated by removing the whey.

Kashk Dried non-rennet cheese made from either dried fermented milk (yoghurt) or dried fermented de-fatted milk (skimmed milk and buttermilk).

Kishk Alternative term: '*tarhana*'.

Kneading Method of mixing aimed at the improvement of the structure (e.g. formation of gluten network in dough; consolidation of the fat phase in butter).

Labneh Non-rennet cheese made from yoghurt by removing the whey.

Lactic acid fermentation Metabolic pathway that releases stored energy by converting organic compounds (such as sugar) into compounds with less stored energy, including lactic acid.

Meal (1) Food eaten together at a particular part of the day. (2) Pulverized whole grain, thus including fragments of bran and germ. See also: flour.

Milk–cereal product Alternative term: '*cereal–milk product*'.

Milk product Any product made from raw milk by clumping of fat, fermentation of carbohydrates, coagulation of proteins, and/or removal of water. A milk product containing processed cereal grains is considered here as a separate food product (see: cereal–milk product).

Milling Preferred term: '*grinding*'.

Mould A container that can be filled with food to obtain a standard size and shape, and to isolate the food from fuel and ash.

Naked (in cereals) Cereal in which the grain kernels are loosely enclosed by their hulls (lemma; palea; and, depending on the genus, also the glumes). The threshing unit is the grain kernel. Alternative but not preferred term: '*free-threshing*'.

Non-rennet cheese Cheese in which the coagulation of proteins is caused only by lactic acid fermentation. Alternative term: '*acid-coagulated cheese*'. See also: kariesh, chakka, and labneh.

Noodles See pasta.

Ordinary butter Butter obtained after churning. The shelf-life is limited due to the presence of some water with dissolved and soluble nutrients.

Pan oven A single-chambered or double-chambered oven with a pan on top that is directly heated from below.

Par-boiling A treatment of rachillate spikelets prior to de-graining that involves soaking and heating by boiling or steaming.

Parching Dry-heating. Preferred term: '*roasting*'.

Pasta Pieces of unleavened dough made from a mixture of meal or flour, water, and salt. The (final) heating is done by boiling. Similar product: noodles.

Pearling Continuously separating the hulls, the bran, and the germ from the spikelets or florets (Delcour & Hoseney 2013). General sense: the processing of the florets of hulled barley, in which the hulls are tightly connected to the fruit wall. Strict sense: the removal of the bran and germ from the grain kernel of de-hulled barley or naked barley (Baik 2014), which is synonymous with polishing. Here, we use the definition applied to florets.

Pearling product Mixture of plant parts produced by pearling. Such a mixture is separated into a principal product (termed pearling unit) and a by-product (termed pearling remains).

Pearling remains By-product of pearling, consisting of hulls, bran, and germs.

Pearling unit Principal product of pearling that is obtained after separation. A pearled grain kernel consists of starchy endosperm. In pearled barley, the part of the palea located in the ventral groove can still be present.

Peel oven A single-chambered, pre-heated oven or a double-chambered oven that is either pre-heated or direct-heated. Bread is baked without a mould or in a mould. The bread or mould is put on a horizontal plate with an implement consisting of a handle and paddle-like head, known as a peel.

Pericarp Fruit wall, which has been developed from the ovary wall after fertilization. Depending on the fruit type, distinct layers can be separated.

Plate oven A single-chambered or double-chambered oven with a plate on top that is direct-heated from below.

Polishing Separating the bran and the germ from the grain kernels, applied to hulled cereals after de-graining and to naked cereals after threshing. See also: whitening. Alternative but not preferred term: '*de-corticating*'.

Polishing product Mixture of plant parts produced by polishing. Such a mixture is separated into a principal product (termed polishing unit) and a by-product (termed polishing remains).

Polishing remains By-product of polishing, consisting of bran and germs.

Polishing unit Principal product of polishing that is obtained after separation. A polished grain kernel consists of starchy endosperm.

Popping Sudden expansion of seeds or fruits as a result of heating, by which the starch swells. Alternative term: '*puffing*'.

Pounding Fragmenting with a vertical movement that is usually done with a mortar and pestle.

Pressing A process that (1) combines fragmentation and separation and is used for the extraction of liquids or (2) reshapes food, for example, by flattening grain kernels into flakes.

Primary product Alternative term: '*principal product*'.

Principal product In cereal processing: post-harvesting product of primary interest produced at different stages of fragmentation (in cereals: threshing, de-floretting, de-graining, pearling, and polishing) and subsequent separation. In cereal processing there are two possibilities: (1) the grain, in which culms and chaff are the by-product or (2) the culms, in which the grain is the by-product. In milk processing: butter and cheese. Alternative term: '*primary product*'.

Puffing Alternative term: '*popping*'.

Pulverizing Fragmentation into powdery fragments (such as meal and flour).

Qara qurut Solid or dry whey. If water is removed by heating, its colour ranges from brown to black.

Rachilla Main axis in a spikelet.

Rachillate spikelet Combination of a single spikelet and a rachis segment. The diaspore of wild wheat and the threshing unit of hulled domesticated wheat is a rachillate spikelet.

Rachillate triplet Combination of three spikelets and a rachis segment. The diaspore of wild barley is a rachillate triplet.

Rachis Central or main axis of the inflorescence and the infructescence.

Rachis brittleness Difference in rachis toughness. In barley, two kinds of brittleness occur: brittle (wild barley) and non-brittle (domesticated barley). In wheat, three kinds occur: brittle (wild wheat), semi-brittle (domesticated hulled wheat), and non-brittle (domesticated naked wheat). Fragmentation of the brittle rachis is the result of the ripening of the infructescence, whereas fragmentation of the semi-brittle and non-brittle rachis can only happen as the result of threshing.

Rachis fragment (of infructescence) Fragment of the rachis consisting of (a) a single rachis node or a rachis internode, (b) a rachis segment, or (c) longer parts of the rachis. The lowest rachis segment(s) can remain attached to the upper part of the culm.

Rachis internode (of infructescence) Flattened part of the rachis between two successive rachis nodes.

Rachis node (of infructescence) Thickened part of the rachis where spikelets arise. This part of the rachis can be used for quantifying the threshing remains.

Rachis segment (of infructescence) Combination of a single rachis node and accompanying rachis internode.

Raw milk Unprocessed milk.

Rennet Extract of the inner part of the stomach of a young mammal (usually the fourth stomach [abomasum] of a calf) that contains rennin.

Rennet cheese Cheese in which rennet is used for the coagulation of casein proteins. Usually, lactic acid fermentation also takes place and this contributes to the coagulation of casein proteins as well.

Rennin Enzyme secreted by the stomach of young mammals that hydrolyses the soluble protein caseinogen into casein. The casein forms insoluble calcium–casein compounds, known as 'curds'.

Rhachis Alternative spelling of '*rachis*'.

Roasting Cooking technique that makes use of dry-heating in which heat transfer by radiation predominates. Alternative term: '*parching*'.

Rusk bread Twice-baked bread that is cut or sliced before the second baking.

Second threshing We advocate avoiding the term second threshing because it is unclear. This term is unclear because it merely refers to a second step in fragmentation, without taking into account the difference between de-floretting, de-graining, pearling, and polishing.

Secondary product Alternative term: '*by-product*'.

Seed *Sensu stricto*: a ripe seed bud. In angiosperms, the seed develops within a fruit. *Sensu lato*: a fruit or a fruit covered by an exocarp, in compound concepts (e.g. seed bank, seed dispersal, seed yield, seed loss). Although the term diaspore is a morphologically viable alternative, the term seed is considered a conserved term in such compound concepts.

Seedcoat The wall around a seed. Latin term: '*testa*'.

Seed dispersal General term for the dispersal of the seed irrespective of additional structures (fruit and exocarp) that may be part of the dispersal unit.

Skimmed milk Liquid, de-fatted by-product of cream making.

Skimming Separating the cream from the now-skimmed milk by removing the cream.

Smoking Heating with smoke, resulting in the removal of water and the addition of flavours of the fuel.

Solid milk product Milk product with a firm, but still compressible, structure. Examples are butter and some cheeses. See also: dried milk product.

Sour milk Fermented milk. Term used to indicate that milk has become acid due to fermentation. See also: yoghurt.

Spike Inflorescence (flowering stage) or infructescence (fruiting stage), in which sessile spikelets are attached to a central axis (rachis).

Spikelet Unit consisting of one or two glumes and one or more florets. See also: rachillate triplet and rachillate spikelet.

Sprouting Alterative term: '*germinating*'.

Starchy endosperm Inner part of the endosperm, thus without the aleurone layer, containing much starch.

Steaming Wet-heating by water vapour.

Strained fermented milk Milk that has become sour due to lactic acid fermentation and has become (semi-)solid by the removal of water with dissolved and soluble nutrients.

Straw Part of the threshing remains that consists of the culm and leaves.

Tabun A single-chambered oven that is heated with solid fuel from the outside. Bread is baked in a horizontal position.

Tannur A single-chambered oven that is heated from the inside and pre-heated when solid fuel is used. Bread is baked in a vertical position.

Tarhana Dried cereal–milk product. It is an intermediate food made from whole or fragmented grain mixed with unfermented milk or fermented milk (yoghurt or buttermilk). Other ingredients may include salt, yeast, vegetables, herbs, and spices. Par-heating can be applied to the cereals and/or the cereal–milk mixture. Fermentation occurs in the milk product and/or the cereal–milk mixture.

Testa Latin term for '*seedcoat*'.

Thrashing Preferred spelling: '*threshing*'.

Threshing Fragmentation of the culm and infructescence in cereals. Fragmentation is followed by separation. The principal product is a floret (hulled barley), a rachillate spikelet (hulled wheat), or a grain kernel (naked barley and naked wheat).

Threshing product Mixture of plant parts produced by threshing. Such a mixture is separated into a principal product (termed threshing unit) and a by-product (termed threshing remains).

Threshing remains By-product of threshing that is obtained after separation.

Threshing unit Principal product of threshing that is obtained after separation. In barley and wheat, three kinds of threshing units are recognized: (1) floret of hulled barley; (2) rachillate spikelet of hulled wheat; (3) grain kernel of naked barley and naked wheat.

Toasting Alternative spelling: '*roasting*'.

Vaporization Separation method in which water is removed from the liquid at the surface (evaporation; sun-drying, or air-drying) or by the formation of water gas throughout the liquid, which escapes at the surface (boiling).

Whey Liquid part that is separated from the solid curds. It is considered as a by-product in cheese making. Sweet whey is separated from unfermented milk (by curdling with rennet), and acid (also termed sour) whey is separated from fermented milk.

Whitening Separating the bran and germ from the grain kernels. The term whitening refers to the white colour of the end product. Preferred term: '*polishing*'.

Whole flour Alternative term: '*meal*'.

Yield The part of the harvest that has economic value.

Yoghurt Fermented milk. A milk product that has become semi-solid due to the curdling of casein proteins, which is caused by the low acidity.

Yoghurt cheese See: labneh.

Indices

Subject index

Taxonomic index